The Managed
Services Playbook

The Managed Services Playbook

*A Guide to Running Successful Managed
Services and Cloud Businesses*

Ed Nalbandian

THE MANAGED SERVICES PLAYBOOK
A Guide to Running Successful Managed Services and Cloud Businesses

iUniverse books may be ordered through booksellers or by contacting:

iUniverse LLC
1663 Liberty Drive
Bloomington, IN 47403
www.iuniverse.com
1-800-Authors (1-800-288-4677)

ISBN: 978-1-4917-3363-9 (sc)
ISBN: 978-1-4917-3364-6 (hc)

Library of Congress Control Number: 2014908600

Printed in the United States of America.

iUniverse rev. date: 08/22/2014

Table of Contents

Acknowledgements vii

Chapter 1 - Introduction.. 1

Chapter 2 - Overview... 6

Chapter 3 - The Offer 10

Chapter 4 - Business Development 37

Chapter 5 - Service Delivery.. 82

Chapter 6 - Client Management 123

Chapter 7 - Strategic Operating Plan 132

Chapter 8 - Operational Management.................................... 147

Chapter 9 - Reporting 155

Chapter 10 - The Managed Services Inspection 163

Chapter 11 - Valuing a Managed Services Business 179

Chapter 12 - Looking Back.. 186

Chapter 13 - Looking Ahead.. 193

Chapter 14 - The Agony of Managed Services 201

Chapter 15 - Managed Services Ten (plus one)
            Commandments 205

Chapter 16 - Concluding Thoughts 207

About the Author.. 209

Acknowledgements

There is a long list of people I have worked with over the years who have helped shape and teach me in this world of managed services, and a number I would like to thank for their contributions to this book. Whenever I say "I" in this book, it really means "we," the collective managed services teams I have worked with at various companies I have been a part of over the years. I'd like to in particular thank Bill Strain, Charlie Mantione, Glenn Jenkins, Shane Sullivan, Craig Grills, Irina Boyeva-Jacobs and Phil Swindle for their help in creating this book.

I'd also like to extend my thanks to my family for putting up with my sometimes crazy work habits: my wonderful wife, Tracy, and my two amazing kids, Kellie and Eric. No one could be luckier to have a more supportive family, friends and co-workers to support my managed services addiction!

Chapter 1 - Introduction

"Managed Services is the hot new trend in the world of IT" - a phrase I have heard on and off for the past thirty years. Certainly IBM felt that way in 1982 when it spun off a new independent business unit called the IBM Information Network. I've spent the past thirty years since I started as a sales trainee for the IBM Information Network in its first year of existence trying to figure out how to succeed in managed services. I'm not exactly sure how I've spent the past thirty years in the same line of business, it's not something I am proud of, but that's what I've been doing. I've watched managed services go in and out of style, and I've always been fascinated by the business - how to generate and deliver a value proposition for clients that is competitively differentiated and run that business in a growth mode that generates strong profit for me as the vendor and delivers the cost savings and performance the client is looking for. Along the way, in trying to figure that out, I've always been curious how others are *really* doing it - what insights do they have that I could use? Am I doing the right things to be as effective and efficient as I can be? There never was a place or resource to go to where I could get that kind of insight. And that really is the inspiration for this book.

That's why I call this the Managed Services Playbook. It's the "plays" I have run over the years in all parts of running a managed services business. It really should be called "a" playbook vs. "the" playbook. This book does not represent how you have to run this type of business, it represents how one person has done it. I am hopeful this playbook will be helpful to those that are a part of running a managed business today or are looking to in the future. I have run managed services businesses that had a few million dollars in revenue to those over a billion. The principals in this book I have found to be helpful in all sizes of managed businesses. The similarities far outweigh the differences in running a large and small managed business.

For the purposes of this book, I will define managed services as when a client pays a vendor to take over or complement their support team in running a part, or all, of their IT infrastructure. There are basically two versions of managed services, outtasking and outsourcing. Outtasking is where the vendor is really complementing and helping a client get optimal performance out of a part of their IT infrastructure. Outsourcing is where the vendor is essentially running a part of the IT infrastructure for the client, typically with little client involvement. The assets a vendor is managing can be on the client's site or hosted centrally by the vendor.

There is a proactive, typically remote, element in managed services, performing at least some amount of proactive monitoring of the client's infrastructure. A recurring revenue stream does not necessarily mean it's a managed service. Having dedicated people on site (or off) and using the client's tools to manage their infrastructure I think of as staff augmentation vs. managed services. The line between basic break-fix maintenance and managed services is typically the element of proactive monitoring, when you proactively monitor you have crossed the line into managed services. Doing MACD's (move, add, change, deletes) and/ or having dedicated people on site helping the client on top of maintenance can be a gray area between managed services and maintenance. I think the true dividing line when you cross into managed services is that element of proactive monitoring. On top of monitoring is typically operating and optimizing a client's infrastructure. It may or may not include transforming the client's infrastructure. The actual technology you are supporting as the managed vendor may be on the client's site or hosted centrally in your or a third party data center. In general, the client in managed services looks to the vendor as "owning" a part of their infrastructure in terms of performance and support - and the vendor must be ready and willing to step up with the expectations that implies.

There are many aspects of managed services above proactive monitoring which are well defined via ITIL (Information Technology Infrastructure Library), which we will review later. These include areas such as configuration management, incident management, service desk, problem management and change management, among others. So while there are many definitions of managed services, the above is what I will use for purposes of this book.

I am hopeful there are nuggets that anyone involved in a managed services business can get out of the book. For those with significant experience in this area, some of this may be a bit elementary and obvious, but I do feel some of the plays and finer points may be beneficial. For those who are new to this area, some of this may be way too detailed in parts, but I feel as you grow and get more involved in managed services, even the more detailed sections of the book may be valuable over time.

This book is not about the macro trends that are driving more and more companies to look for managed services from their trusted vendors. It's not going to explain why

IT infrastructure vendors of all types should be investing in managed services. Likewise, this book is not oriented around how to start a managed services business or move an existing maintenance or professional services business to a managed one. The book is geared towards vendors that have an existing managed services business. My assumption, as you begin your investment of time in reading this book, is that you are aware of <u>what</u> the trends are that are driving and leading to unprecedented growth in managed services and <u>why</u> IT vendors must have competitive managed capabilities. This is about <u>how</u> you do it. How you can create and grow a profitable managed business from someone who has done it for a long time, made many mistakes, had some success and remembers enough of both and decided to write it all down before he forgets it all! This is basically an "on the playing field" guide to help your team win.

You have to be "all in" to be successful in managed services. It is a niche area with very specific nuances that must be navigated with forethought. It's not something that can be picked up in six months, mastered in a year or two, then move on to another IT area. For me, for some reason, this business area has always fascinated me. As I've often said, it's not that hard to build and grow a managed services business, and once you have a base of clients, it's not that difficult to drive profit out of the revenue stream you have produced. However, having both an ongoing, high growth managed services business, with increasing and strong profit margins, is one of, if not the, toughest thing to do as an IT vendor. When you get there, the value that business brings to shareholders and clients will last a generation. For thirty years getting a managed business to that state has been my passion. I've addressed this market as a part of a large integrator, a service provider, an equipment manufacturer, an offshore provider and an independent start-up. And here is the playbook I've used to get there. I'm sure there are many ways to get there, and what follows is one of them.

Chapter 2 - Overview

The current era of technological change is certainly dizzying, social media, as well as mobile, tablet and hybrid devices, are all blurring the line of consumer and business technology use. We are seeing successful technology IPO's and a significant influx of venture money into tech companies. One constant amongst the ongoing rapid technology changes is businesses continually looking to adopt and adapt technology to improve competitiveness and productivity. However, the technology downturn over a decade ago has had an indelible impact on how companies plan and deploy technology. There is an innate conservatism on new technology purchases, as well as a focus on flexible purchasing models vs. large capital outlays. The book *Consumption Economics* by J.B. Wood, Todd Hewlin and Thomas Lah does a really good job summarizing the trend towards this new utility model that has emerged. And Managed Services is at the heart of this new trend.

Managed Services is not sexy, it never really was nor will it ever be. According to IDC, it's a 9.9 billion dollar per year industry growing at around eight percent. Its growth will never be as high as the rest of technology in boom periods, and it won't decline as much in economic downturns. At its core, Managed Services is still the nuts and bolts of helping companies manage their infrastructure, however it's now evolving fairly dramatically. New consumption models where the vendor holds the asset, including public cloud, private cloud and hybrid models, are an irreversible trend. The basics of proactively managing a company's infrastructure remains as these models continue to evolve and expand. "Cloud" in its various forms gets all the buzz now, but underneath the buzz is the management of client IT infrastructures. As a result, IT vendors of all types and sizes are looking to perfect and optimize their managed services capabilities as these client consumption models continue to get better defined. For as companies everywhere are realizing, it's not just about deploying new technology, it's about how to truly drive business impact from that technology in the daily operation and integration into the business. The "serviceability" of any new technology is now at the forefront of CIO's minds, and hence the growing importance of managed services.

As we find ourselves in the midst of a new tech boom, it's a boom fueled mostly by consumers vs. businesses. New technology spending by companies is not manifesting itself in growth for most IT vendors. How to succeed in this new technology consumption era is still a work in progress for most IT vendors. Seventy nine percent of the Technology Services Industry Association (TSIA) Services 50 (comprised of fifty of the largest global technology companies that publically report on product and services revenue) reported flat or declining revenue from 2012 to 2013. In fact, for that same Services 50, their product revenue has declined by 48 billion dollars from Q1 2011 to Q1 2014, while their services revenue has improved by 12 billion dollars from Q1 2011 to Q1 2014. With professional services, support services and education services roughly flat over that time period, managed services is the only business growing on the services side. Most of these vendors are working feverously to change their models to meet new client desires on how they want to procure and use technology. The importance of managed services really cannot be overstated, it truly is at the forefront of the move to this new utility model. How IT services vendors can best build and run these parts of their business is the goal of this book.

This book is targeted towards those of you who already have a managed services business. Hopefully it's up and running and going well. You are acutely focused on maximizing sales and efficiency, as you look to drive revenue growth and profit in the near and long term. It's an all-consuming effort that can sometimes find you so focused on tackling opportunities and issues that you don't have the time to step back and assess am I doing all the right things? What is my competition doing that I may not be doing? My hope is after reading this book you will not only get some good ideas and "plays" to improve your business, but you will also gain more confidence as you improve the key areas of your business that your plans are the right ones and you can fully focus on execution.

The content of this book is basically a look at each major part of a managed services business and outlining best practices on how to be successful. What are the plays that work well in all areas of a managed business, how you can best manage and inspect the business and what to look for in evaluating how you are doing are the key questions this book attempts to answer.

The first half of the book focuses on four main areas of a managed business. It starts off with the offer and reviews managed services offer definition, pricing, strategic planning, differentiation, value proposition and marketing. The next chapter is on managed services sales and business development. This chapter reviews the team and metrics, your game plan and approach to selling, the actual client sale, and the contract. A review of service delivery follows in the next chapter, which includes core operations support, platform, process, Service Level Agreements (SLA's) and metrics. The next chapter reviews client management, and what the key actions that are needed to drive successful satisfaction, retention and upsell.

The second half of the book focuses on managing the business on an overall basis. There are chapters on creating a strategic operating plan, operational management, reporting, an inspection chapter to evaluate how you are doing in achieving your goals, and a chapter on valuing a managed business. The final few chapters include a short look back at the managed services industry in the past, a look forward at where it is going, a chapter on the 'agony' of managed services and we finish up with a review of the managed services ten commandments (plus one).

I hope this comprehensive look at the different parts of a managed services business and how to best operate and run it will provide value to those in the middle of the fight. While there may be certain parts of the book you may want to jump to first, I would recommend reading from the beginning, as I believe all the items covered here are interrelated to a fair degree.

Now let's attack each part of a managed services business to insure you have the plays to create a winning game plan and have it run in an optimal fashion!

Chapter 3 - The Offer

If you're already in the managed services business, you have a delivery capability built with a platform and processes in place, an offer your sales team is selling and a business plan behind what you are doing. Each section of this book will outline from my experience the "what" and "how's" to make each part of the business a success. And this first section will deal with your managed services offer. The offer management (aka product line management) function is a critical component of every managed services organization, providing the linkage between what clients are looking for and the delivery of service. I will try to focus here on what the keys are to successful offer management in managed services vs. offer management in general. This section will cover ***strategic planning, offer definition, pricing, differentiation, value proposition, marketing and breadth of responsibility***.

Strategic Planning

I do believe every managed services organization each year should do a strategic plan, a "full circle" plan for the business. I believe the offer management team should drive this. I personally don't like having a separate strategy team or person to drive this. It's not only the key strategic actions in each part of the business that need to be part of the strategic plan, the financial plan and key performance metrics should be integral components as well.

A thorough understanding of the *market size and market opportunity* is imperative and not always easy to get an accurate read on. The analyst community that truly understands and covers the managed services market is small and hard to find. You do not find the best analyst for the type of IT managed service you are providing by looking at different analyst companies, it's the individual analyst that knows that particular market which you have to find. If you don't know an analyst you have confidence in, a process of meeting with and "interviewing" analysts from the different industry analyst companies is a must. I won't give a rundown of the analysts I have found to be helpful or not so helpful, but there has been one I have worked with for over ten years, Eric Goodness of Gartner Group. He has an excellent understanding of the managed services market. His insights were invaluable to me in each company I was in. Reviewing your strategic plan with an analyst before it's done, as you are forming it, I find can be very helpful. Ideally you want an analyst to help you shape it, not tell you how good or bad is after the fact.

I mentioned TSIA previously. They have a relatively new practice around managed services which I would recommend to any managed services vendor. They have a tremendous amount of data on managed services organization structures, delivery processes, sales and go to market structures, and P&L governance practices, among other areas. They also have excellent data on managed services metrics. I am a founding board member of the TSIA managed services practice and have found it very valuable.

In my role at Avaya, our strategic plan started with a market analysis that showed communications managed services growing at about 5%. From market intelligence, we knew what Cisco and Siemens were growing at and targeted a growth rate higher than them. The growth rate of your target market should be a key factor in determining your revenue growth goals.

It's important to gain a thorough understanding of *client requirements* (including where the market is going) and your *competitor's offerings*. Analyst reports, where available, can help, but it's really from the front line where you get the best information. Your sales force and the clients they talk to, of course, are a very important source of information. In general, it is not too difficult to get a good handle on the client needs of the part of the market you are going after. Truly understanding your key competitors' offerings can be difficult, but it's of utmost importance. Client bids where you have competed and gained an understanding of your competitor's offerings is important to note. Certainly you should be able to talk to analysts who can help as well. One thing I have found effective is to work with a consultant who can go to a competitor for a "blind bid" on a set of services you outline. It's easy to find consultants who can effectively put together a blind bid.

Among the areas you can compare yourself to your competition include remote management capabilities, on-site support, offshore capabilities, consulting skills, productized offer attractiveness, custom capabilities, deal flexibility, price competitiveness, technology breadth supported, ability to include assets as part of your solution (on your books), and strengths in enterprise/mid-market/SMB.

Another key part of the strategic plan is to identify up front the target markets you are going after and not going after. I think it's important for anyone connected with your managed services business to understand that. Figure 1 shows one way to identify target markets. I have always struggled with segmenting along industry lines, as managed services tends to be more "horizontal" vs. "vertical," but that is another way to do it. When you have true consulting as part of your value proposition in managed services, industry segmentation is more appropriate. Also, when you get into application layer management, industry segmentation is often critical.

Figure 1:

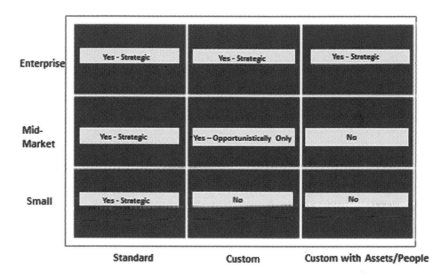

Your analysis of client requirements and the competitive landscape in your area, along with your offer definition and value proposition, should lead to a clear outline of your differentiation in the market. The key to the strategic plan should be to show there is a growing market you are going after, with an offer set that meets what clients are looking for, in a competitively differentiated fashion.

The rest of the strategic plan should outline your game plan from a business development, offer, delivery, tools and client management perspective. Outlining what you plan to do in each area, with key actions and owners along with metrics, should be the bulk of the review. We'll discuss metrics in each chapter of this book. I believe in having a few key metrics in each part of the business that are religiously tracked.

The final section I typically have in the strategic plan document is the *financial plan*. In addition to a 3-5 year high level view of your financials, it should include a detailed analysis of the upcoming year. The revenue should be laid out by offer and geography along with a review of your base revenue and required new revenue (which in turn translates to your bookings objective). The cost and associated gross margin for each offer should be documented, along with your investment plan. Chapter 7 will review in detail the financial management of a managed services business. It should all come together in a clear, comprehensive view of the financials in the strategy

Offer Definition

In effect, this is where it all starts - the fundamentals of describing the features and functions you will offer at what price. Your offer starts with the overall plan for your managed services business, what your fundamental value proposition is and associated technologies you will support. How you define the support you will provide into an offer definition is a key part of having a successful business. The beginning of this process, in my mind, is outlining your offer along ITIL (IT Infrastructure Library) lines. For those who may not be familiar with ITIL, it's basically a set of definitions for managing IT infrastructure. Created in Great Britain, it has become the de facto standard and 'common language' in managed services. Many clients and most IT providers are familiar with and use ITIL, at least partially, to run their IT infrastructure. Figure 2 is an overview of ITIL and the different functional aspects of managed services it defines. While it is easy to find flaws in how the functions are defined in ITIL, I believe it's not worth the effort to find the flaws. The benefits externally and internally of using ITIL given its broad acceptance outweigh any imperfections in its definitions. I've resisted at times using ITIL fully, and in the end it was mistake. Your standard offers, as well as custom client SOW's (statements of work), should be structured along ITIL lines. The critical communication between offer management and delivery internally, in terms of what will be provided and not provided, is made significantly easier with this common language.

Figure 2:

ITIL Overview

Your offer definition is truly the key building block for how your managed services business is run. A well-defined offer should be the epicenter of how you drive efficiency and value into your business. I'm not going to be overly prescriptive in suggesting how best to define your offer, each provider's offer will vary greatly depending on the part of IT you are managing and the market you are targeting. Of course, the most typical method for structuring managed services offers is a "bronze, silver, gold" - like format, and it's certainly typical of most providers. I prefer having a defined standard offer as the "silver", and then having an "express" or "lite" option as the "bronze" and a premium option as the "gold". As you move up your "offer stack," the offer builds up in functionality, that is, the standard offer builds on the express offer and the premium builds on the standard offer. Figure 3 is a general way you could potentially breakout your offer.

Figure 3:

Premium Offer

- Capacity Management
- Availability Management
- Change Management
- Release Management
- ➤ Includes Standard capabilities

Standard Offer

- Service Management
- Incident Management
- Problem Management
- Configuration Management
- Service Desk
- ➤ Includes Express capabilities

Express Offer

- Service Maintenance
- Client Notification
- Proactive Monitoring
- Parts, Patch Dispatch (maintenance)

Of course, you can come up with better names for the offers than I have here. And there is no magic to which functionality you put where. I do believe it's worth some thought on how you can structure your offers so your clients don't automatically default to the lowest price option, which sometimes happens in a gold, silver, bronze format. That's why I like having a "lite" or "express" option, which can help to push many clients to your "standard" or silver offer.

The offer structure should be based on the value you provide, with a differentiation of that value coming between offerings. As you move up the stack, you ultimately build an efficient way for a client to entrust the operations of their environment to you. The break points between the different levels of your offer create the ability for the client to integrate the delivery of specific tasks into their operations.

These offers, and the ITIL-based categories underneath them, should form the basis of how your organization is run. It should impact how service delivery is organized and should be how your revenue and costs are tracked. One decision each managed services organization must make is around service level agreements (SLA's). Are you going to have them in client agreements, will you entertain financial remediation, what SLA's, if anything, will you have as part of your standard offer?

One thing is for sure, you should have a strategy around SLA's and how you plan to handle it as a business. It's critical to have a small "center of excellence" in delivery (which could be one person) who is responsible for SLA definition and management. In larger deals, it will almost always come up and can extend sales cycle times. In general, I believe in proactively proposing SLA's with remediation, especially in larger deals. I actually find clients are generally reasonable, especially when you don't show resistance to the concept. They simply want you to feel pain, and are incented to avoid, service impacting issues. Another key point for larger deals is insuring you get a good baseline of the client environment against the SLA's you are negotiating. You need to insure you can hit those SLA's as is, and if not, there needs to be a plan to get there or appropriate language must be included in the contract. We'll discuss the whole SLA area in more detail later.

Pricing

The pricing of your offers is certainly a key area for both your offer management and business development teams. It starts with the prices for your standard offers outlined above. Standard offer prices are typically a monthly unit price for the part of the infrastructure you are managing. Most typical is per element pricing, where you have a price per month by the type of device you are managing. It could be a per router price for data, per port for voice, per server for server management, per mailbox for messaging, etc. Many managed services vendors have moved to per user pricing, where the user can have multiple devices. I don't have a huge preference as a vendor and have not seen clients push hard one way or the other. I would suggest using whatever you can track and manage most effectively and is relevant to the offer and client. The past few managed businesses I have run we have used both. I have also seen some vendors employ a fixed fee for all services you are providing, and some I've seen using a "maximum" fee. To me, the risks of undercharging here outweigh the benefits of increased client certainty in their costs. I would rather do something like that on an exception basis for clients who ask for it. If you have a client who has relatively static volumes, be it users or devices or whatever, you in effect will have a fixed price for the client.

You will also need to determine if you want to have an upfront one time charge. Clients tend to negotiate out any one time charges, so you need to be prepared to have these charges waived and captured in your list monthly prices or discounting strategy. I have found most often it makes sense to have a set of one time charges that you can be prepared to waive for deals of a certain size.

You also need to account for MACD (move, add, change delete) charges. You can have a separate charge for any MACD work, but I have found it most typical to include soft MACD's in your pricing, at least up to some volume level per month, and then charge separately for 'hard' MACD's. A hard MACD involves an actual physical change of a piece of the client's IT infrastructure vs. a soft MACD, which can be done online remotely. Chapter 7 goes into some sample pricing and how P&L management plays a key role in how you price.

Once you have the structure of how you will price in place, it is important to get accurate benchmarking on what pricing your competitors have for a similar service. You can look to analyst reports to help in many cases, such as those from Gartner Group and others. Sometimes I have found these reports to be on the high side, so you have to make sure you don't get a false sense of security around how you are pricing. Vendors often give analysts their list prices vs. actual "street" prices. The best information is often gleaned from the managed services deal teams working deals, understanding what competitors bid for similar services. Many companies will give you a copy of a competitor's bid if asked. As mentioned previously, you can also work with an analyst firm get a "blind bid" from your competitors. You can get a pretty good handle on their pricing that way. Another way to get a good handle on your competition's pricing is to hire folks from them, something I have frequently done. What's most critical in comparing pricing is to understand the specific services being delivered for the price, which is why actual competitive bid situations or a doing a "blind bid" typically gives you better competitive pricing data than an analyst report.

Large custom deals often are the deals that make or break managed services businesses. So it follows that pricing of these deals is extremely important. As with other aspects of custom deals, you want to begin your pricing with the standard pricing. The core of your delivery should be based on your standard offers. The delivery deal team involved will need to cost out all non-standard functions you need to support for any given client. As with all aspects of custom deal development, the business development team must work closely with the delivery team on this costing exercise, making sure reasonable assumptions are being made. We will discuss more later, but it is critical that the delivery team be active participants in large, custom deal proposals. It is not something to shy away from, fearing delivery folks could extend or jeopardize the close. Done correctly, it should improve the client's confidence in your understanding of their environment and your ability to deliver. It will also increase the odds of a successful client relationship (and upsell) greatly.

It is important to have a pricing lead for your managed services business that sits in finance, offer management or in business development. That person will work with the offer mangers to determine standard prices, as well as work on a deal by deal basis with the bid development team to determine the pricing of each custom deal. A proforma should be developed for each custom deal, looking at the revenue, profit and cash flow over the term of the contract. A predefined bid approval process should include financial approval of all custom deals.

The pricing of your managed services business is serious business. We will review in a great deal of detail the implications of pricing on the overall health of your business, and how it's a key piece of your financial model and plan, in Chapter 7. All decisions on pricing for standard and custom deals should be made in the context of your overall P&L, so having someone assigned to lead pricing that fully understands all appropriate details is critical.

Differentiation

To me, the epicenter of any managed services business is your competitive differentiation. It's imperative to have a complete obsession with what makes your solution different from and better than the alternatives available to your clients. Too often I see managed services businesses that focus on what they are good at doing and what they have built as their differentiation. Just because you built something or have a core competency in something doesn't mean you're offering something that is tangibly different from your competitors.

It's important to be fairly granular in the assessment of your differentiation. For example, saying you have global support is too high level. You really have to identify what it is precisely in providing global support that makes you different or better. Aligning your offer along ITIL lines isn't a differentiator. Pushing yourself to have tangible differentiating qualities is critical, and it should lead to areas you want to invest further in.

There are so many aspects to providing IT managed services and different target markets, it is hard to outline exactly what key differentiation you should be pursuing. I'll provide two examples that may be helpful.

At Avaya, the managed services business traditionally provided a level of managed support slightly above basic maintenance for Avaya technology around Unified Communications (UC) and Contact Center (CC). It was a fairly ordinary offer with below ordinary revenue growth, to put it mildly. There were many initiatives we undertook to improve the offer, most of which were not differentiating the offer, just getting it up to par. The business took off as we identified and invested in areas that made us different than our key competitors, Cisco and Siemens. Without going through too much detail, here are a few of the key differentiators we had that propelled the business from a declining revenue trajectory to one that began to grow at a 25%+ CAGR.

> **Multi-vendor support** - communications clients that had multi-vendor environments had a very difficult time getting a managed services provider to support them. Cisco was Cisco only, Siemens did some multi-vendor, but were very poor in Avaya support, and System Integrators and Service Providers (SI and SP's) were overwhelmingly Cisco-only focused. We invested from a tools and engineering standpoint to support virtually any communications vendor. This became a clear differentiator for us.

> **Utility (Opex) Private Cloud Offer** - Avaya found a niche where clients were looking for a utility offer (one unit price inclusive of hardware, deployment, and

day 2 support) in a private cloud environment. Public cloud offerings were not tailored to their specific needs, yet the notion of not spending capital and having the flexibility to ramp up and down in a utility model was very appealing. This required us to put assets on our books and deliver a true utility model. Cisco did not do this and many/most of their large partners didn't either. Siemens would do it; but approached it in a clumsy fashion. This became a key differentiator for us.

Voice Application Support - Potential communications managed services clients were looking for vendors with deep technical expertise, more than they could provide on their own. We simply found that when it came to UC and CC application-level support, from a quantity and quality standpoint, we had much more of it than Cisco and their key partners.

There were other differentiating elements, such as doing broad global deployments (with multiple vendors), providing custom stringent SLA's with remediation, providing analysis and reporting in a superior fashion to our competitors with the tools we invested in, and our ability to customize an offering to a client's needs better than our competitors. The more we learned about what our clients wanted and what our competitors were truly doing, the more it crystalized what we needed to focus on and invest in. The bottom line was we created an offer construct where clients could get all the commercial benefits of a cloud, opex model, in a customized, best-in-class delivery model. Being able to integrate a technology transformation with a differentiated delivery capability, masking the complexity from the client, and all in an opex model with no up front investment, fueled unprecedented growth for Avaya, as well as benefits for clients that took the journey with us.

Another example of differentiation comes from Cognizant. The focus there was on creating a differentiated value proposition vs. other offshore vendors, such as Wipro, Tata, and Infosys. IT Infrastructure Services is what the offshore players called managed services, and it's typically not much more than a staff augmentation play with T&M (time and materials)-like pricing. In a similar fashion to the application development business that fueled their rapid growth, offshore players apply a set of offshore resources to help a client manage their environment. And they typically use the client's tools, processes, etc. There are typically no SLA's. It's a pure staffing play and that's how it's priced, by person per hour or month. There were several key things Cognizant did that differentiated it from their competitors.

> **Remote Management** - Cognizant had a full remote platform that was far superior to any other offshore vendors, a full client portal and reporting tools that were leading edge. HCL eventually caught and passed Cognizant in this regard, but for a while this was a clear differentiator.

> **Managed Services Offers** - Cognizant went through a painstaking process of "productizing" all its services. Across the IT stack it supported, you could buy the managed service in a unit price fashion. Not a revolutionary concept for most managed services players, but for offshore players it was.

Onshore / Offshore Support - Cognizant had a US-based NOC to complement their offshore NOC's, which provided superior US technical support for clients where that was important.

Looking at Cognizant's value proposition and differentiation in managed services, it's important to put it in context of their overall business as well as their competitors. These businesses are built on a labor arbitrage play. It's a powerful value proposition to reduce headcount costs by half or more for basically the same function. This labor arbitrage play is a big part of managed services, putting tremendous pressure on onshore managed vendors. We'll talk more about offshore support for managed services vendors later. But in the case of Cognizant and its offshore competitors, the notion of having productized, unit priced offers remotely managed with a proprietary management platform anywhere in the world (on or offshore) was (and is) fairly new.

There are also certain parts of IT that the offshore players have pretty good experience in (i.e., helpdesk, IT operations) and parts they are weaker in (i.e., networking, hosting). There are exceptions to this (ex., HCL's networking capabilities, Wipro's Infocrossing acquisition), but that is generally a fair statement. Cognizant pushed hard to have skills across the ITIL stack. It tried to leverage its core strengths as an offshore player with a transparent labor arbitrage play, and add more traditional managed services capabilities. It was a pretty compelling, differentiated play that led to very strong growth in that part of their business. From all I hear, IT Infrastructure Services remains a strong growth business for them.

Cognizant is an interesting company I enjoyed working for. Having a strong onshore presence was a key focus for them. They organized in a "two-in-a-box" model, having an onshore and offshore lead for each service line and vertical. Clients reacted well to it. The thought leadership for the business overwhelmingly came from the offshore lead, with the onshore lead often being more sales focused. I do expect the core business strategy Cognizant has, coupled with their fairly differentiated managed services play, to continue to enable them to grow well in the space. HCL now has a stronger managed services play, especially in networking, but I expect Cognizant to continue to thrive in the managed service space.

Getting that *breakthrough* from an innovation standpoint can be the difference-maker for a managed services business. It could be a platform-related breakthrough, technology support, core processes supported by skilled professionals that are hard to duplicate, or any one of a number of items. The one thing I have found is that the breakthrough you need is not often some incredibly brilliant idea (though that is a great place to start). It could be something your clients are asking for that everyone just assumes you can't do. Whatever germ of an idea you have, it's important to involve your clients early. Don't "invest" too much time and money without a real time feedback vehicle by the people that will really determine the success of any potential breakthrough offer - your clients. If you feel an idea has merit, work with a client or set of clients, even if you have to do things for free at first, get their feedback and adapt. You could set up an ongoing client advisory council to be an integral part of the strategic plan. To me, it is most critical to move from the white board to a real operational solution with client(s) to get the feedback and input you need to truly understand the value of your offer.

A few examples of breakthrough thinking I have experienced in my career may help, and it shows how innocuous that first idea could be. I talked about Avaya's differentiation above in areas such as multi-vendor support, private cloud and application support. Those items started as we were trying to sell a new set of standard managed services. Sales of our standard offers were slow, and along the way we had several clients ask for features we didn't have as part of our standard offer, which we routinely walked away from. One large pharmaceutical company pushed us to expand our offer beyond Avaya support and we figured out how to provide it. A large service provider pushed us to include assets in our bid, so we got that through. Over time we found a niche of providing highly custom support that other providers didn't typically do, which led to a highly differentiated play for us. The differentiation we developed at Avaya really came from our clients, it wasn't developed in some internal conference room or from some offer manager's great idea. We leveraged our standard offer and adapted on top of that to meet our client's needs.

At AimNet Solutions, an independent "pure play" managed services company, one of our goals was to provide our standard offers through system integrators and service providers in a "white label" fashion. We had interest from target vendors, but it never really took off. As we worked with those vendors on why, it was apparent they lacked much of the expertise required to really drive the offers from offer management through their sales force. They continually asked for our help, but no amount of our support seemed to be enough. We finally worked with them to provide a formal managed services support offer where we in effect provided staff augmentation for various key roles in their go to market efforts around managed services. Our ability as a company to help managed services providers perform key go to market and offer management support became a key differentiator that allowed us to sell a ton of white label services, which became the largest part of the AimNet business.

Value Proposition

Another key element in offer management for managed services is identifying your value proposition for clients. Your core differentiation is not, by itself, the value proposition - your core value proposition outlines what benefit the client will see directly by using your offer. Your differentiation is why the client can recognize those benefits more effectively by using your service vs. some other provider.

I have found four key areas that have been typically the basis for the value proposition in the managed services businesses of which I have been a part of:

- Lower Total Cost of Ownership - outlining how your offer can reduce the overall cost of a client running the applicable part of their infrastructure.

- Improved Performance - Enabling the client to achieve higher quality services for their end users leveraging your skills, scale and intellectual property.

- Increased Risk Mitigation - putting the client in a better risk position going forward using your capabilities, especially appropriate if the client is undergoing significant changes.

- Accelerate Transformations - Enabling a client to leverage the benefits of any transformation they are undertaking by using your team and capabilities to drive the change quicker and more effectively.

There are other potential value propositions you can have (ie., increased flexibility, more predictable costs, increased transparency/visibility, etc.), but I have liked to center my managed offers around those four areas.

As you crystalize your value proposition, it's important to link these value statements to the capabilities and differentiation you have - that's when an offer truly becomes a compelling to a client.

Marketing

Marketing managed services is a tricky proposition. The ability to spend money with little bottom line results is more likely to happen in managed services than most other technology-related marketing pursuits. I will confess up front I am not a believer in spending much money on outside marketing efforts around managed services. Whether it's advertising, trade shows, or "call-out" demand generation campaigns, I simply haven't seen the cost be worth the benefit. Though I know some managed businesses who have had success doing these things, I have wasted a lot of money on these types of expenditures in the past, though not in recent years.

Of course, it is very important to work proactively at getting your name and brand around managed services out. I'm not sure I have the best instincts around marketing in general, so you can judge my thoughts below with that in mind. Nevertheless, here is what I have found works for me in marketing managed services.

First off, I do believe it's very important to have very crisp 1-2 page "brochures" on your offerings. I do like to have them in hard-copy as well as soft-copy. A nice color layout and logo is all you need. It's the content that is key, focusing in an "elevator pitch-like" format on the core value proposition and your offer differentiation. Some typical core benefits that could come out in your managed services message we discussed in the value proposition section. And as you do that, to highlight your differentiation is critical. It's also crucial that everyone in your business can clearly articulate, in a consistent fashion, the message around your managed services offerings, and this should come out in your collateral.

Getting the word out on your offer and getting the buzz you desire I believe is best accomplished in a "bottoms-up" fashion. First, I'm a big believer in getting an industry analyst or two to become a close confident and adviser. As I mentioned, it can take some digging to find the right one for your company, but the benefits of finding the right one can be significant. I mentioned Eric Goodness of Gartner Group, who is no doubt "the man" in managed services among analysts and who I would choose for sure, but there are others who are good that you can work with as well. As with working with an industry analyst on anything, engaging them as you formulate your offer and message is key. They can help steer where you go, and of course will feel more of a sense of ownership the earlier you engage and use them.

An incredibly important part of a successful managed services business is having good references. This does not just happen by providing great service; it is something that must be cultivated by your managed services organization. It starts in the sales cycle, whether it's a new managed services sale or an upsell of an existing client. It's important to let the client know you would like to earn their confidence to be a reference. In fact, I actually like to try to put it in the contract. It can be done and it works. Just have a 3-4 sentence canned paragraph you'd like to add. For example, it could be that they agree to do a once a month or quarter reference call to clients, analysts, or industry trade rags, provided you are delivering good service. As you go through the negotiations with the client, where they are pushing you on price or other concessions, push to include the paragraph as part of your proposed compromise. No, it's not truly enforceable, but it works. In addition to this, I really like to incent monetarily your client management function to get clients to agree to become a reference. The incentive doesn't need to be big, but it can work.

It's very important to create a culture in your organization on getting references. There are so many managed services organizations where references are such a problem. A determined, patient, thorough effort to gain references should work in every business. Whoever has the best relationship with the right client executive needs to drive this and ask for the "order."

The ultimate "buzz" vehicle for your managed services offer is an article in a trade rag on your offers. And then making lots of reprints and url links in e-mails can be very effective to send. I have found nothing helps create the external market buzz you want more than this. In my experience, there are several things you can do to help make this happen. First, what not to do is to push to get an interview with a publication believing the message around your capabilities is so compelling it has to be of interest and you spend all your time on that. It's important to come to the table with two things. One is an industry analyst who is willing to talk to the publications. As you describe your service, you can offer to the reporter the opportunity to speak to an industry analyst. They will almost always say yes and it adds so much credibility to the message. Of course, analysts like everyone love to see their name in print, so as long as you have cultivated that analyst relationship, it shouldn't be too difficult. Two, get a client who is willing to be interviewed. This is more difficult as many client corporate communications teams block it. It can make the difference between getting an article or not, so it's worth the effort. And yes, you can get this into a client contract, getting them to agree to a once a year interview. And I wouldn't worry about it being a "no name" client. If the reporter can talk to an analyst and a client, you will greatly enhance the odds of getting an article. Industry publications especially love client-centric pieces, given that is their target audience, so keep that as your top priority. And when you get that article, it can often get you more buzz and marketing benefits than having a huge marketing budget for managed services.

Offer Management Responsibility

One question I have wrestled with over the years is how much responsibility the managed services offer management team should have. The group can become a catch-all group, getting responsibility for much of the "tough stuff" to deal with and manage. The general rule I have used is the more standard managed services you are selling, the broader the responsibility offer management needs to be; the more you are selling customized client arrangements, the narrower. Regardless, the role of the offer management team is very important.

When you are selling more customized deals, you still need to push the organization to have a core set of standard service definitions upon which your custom deals should be built. I do feel that the delivery team must be very active in costing out any custom deal, so the more custom deals you have, the profitability of your business will depend less on standard offer pricing/costing. Conversely, a managed services business built mostly on standard deals puts a great amount of importance and responsibility on offer management on how the offer is defined, priced, and costed out. When you are selling mostly standard offers, you really can't do anything without active involvement of the offer management team.

I prefer the offer management team be very involved in marketing. Any dedicated managed services marketing headcount is best spent on the implementation of the marketing your offers vs. the content and core message. Offer management should be the prime message makers to analysts and the press, with a thorough knowledge of their offer, the market, and competition.

In Sum

How you structure and define your offer and value proposition is a critical component to having a successful managed services business. There are some important areas that should be a part of any company's managed services offer plan.

- A "full circle" strategic plan should be done before detailed offer plans are done; a plan which includes market size and opportunity, client requirements and competitive comparison, target markets, offer definition, value proposition, competitive differentiation and a three year financial plan. Detailed plans for each part of your organization (ie., business development, offer, delivery, platform/tools, client management) should all support the strategic plan of the business.

- The heart and soul of your business is the competitive differentiation of your offer - what makes your offer better than the alternatives your potential clients have. Breakthrough ideas to drive differentiation often come from having a strong set of core services and listening hard and creatively to your clients. The best breakthrough ideas are often right in front of you.

- It's important to have a well-defined standard offer along ITIL lines. Having a strong standard offer is critical, even if you typically do custom deals. In addition, defining your offer via ITIL will aid greatly in communicating and working with clients, as well as across your organization. It's the common language of IT that should be embraced by your managed services business.

- The discipline of having a structured approach to managed services pricing is very important, even if most of your client deals are custom. It is this structured approach that should enable you to more effectively work on custom deals, compare your pricing vs. competition, and provide the foundation for your financial model and plan.

- Successful marketing of managed services is not in proportion to how much you spend. The best managed services marketing involves hard work vs. money spent on advertisements, elaborate trade shows, etc. An approach centered on strong client references, analyst and trade publication awareness can go a long way to creating the buzz you are looking for. All marketing must start with well done, "crisp" messages around your offer, value proposition and differentiation.

Chapter 4 - Business Development

Selling managed services is in some ways typical to any technology sale, but in some ways very different. In fact, I would say it is more different than similar. It's not an easy sale in any way - from generating demand to gaining interest to developing a proposal to closing the deal - it's often a tough and long process. Done correctly, it leads to generational relationship between a client and the managed services vendor that is mutually beneficial; done incorrectly, it leads to a ton of wasted time and effort, a lot of hope, and ultimately disappointment.

In this section I will cover four main areas: the **team** you need to succeed, the managed services sales **game plan,** the actual **sale** event with the client and the **contract**.

The Team

The managed services sales support team is a critical component in having a successful managed services business. Most managed services vendors are selling their services though some front line sales force that's not equipped or trained to sell managed services - their core expertise is selling technology, transport, or some other core vendor offering. There is an art and science to selling managed services through a sales force who fundamentally sells something else.

One thing I believe is a given is that you must have an overlay managed services sales team that works with the front line sales team of your core business. If you are an independent managed services player and you sell through a partner, you have the same dynamic. A managed services sale is fundamentally different than a technology or communications services (transport) sale, and you must have someone talking directly to the client that fully understands your offer and its core value proposition - someone that can talk to the client about their pain points, thoughtfully craft the solution to be proposed, and ultimately close the deal.

This sales support role is truly one of the key success factors in having a successful managed services business. The role can have many titles, I'll call it a managed services business development manager (BDM) for purposes here. A BDM will typically cover a certain geographic area or several branch offices or regions of the front line sales force. The BDM is the epicenter of the managed services sales effort, and how that role should be played will be discussed throughout this chapter. As I mentioned, for an independent, or "pure play", managed services sales team, when you sell through a partner, as many do, I think it's fair to consider your sales team to be an overlay sales support group for the partner's front line sales team.

There must be additional managed services support in selling activities outside the BDM. First and foremost, you need a technical pre-sales engineer to work hand in hand with the BDM. This is a must to help provide technical credibility with the client. Ultimately, the client must be comfortable that their performance requirements can be met, and the pre-sales engineer will play a key role in doing that. In addition, I believe in setting up a managed services sales support line, an "800 #" anyone in the front line sales organization can call with any type of question. This sales support line can also be a first point of contact if you are trying to sell managed services through a partner, to support that partner's front line sales force. For the front line sales force, it can be non-threatening way to get questions answered, and when the managed services BDM is not available, to make sure things progress with the client to get a qualified deal. This team can also proactively call-out to both sales reps and end clients to generate demand. I typically like to have one relatively senior person managing the sales support line, with a few more junior teammates manning the line. You can communicate that the line is open during certain hours and promising one hour call back to sales is important to give the level of confidence required. Another benefit to setting up a sales support line is to continue to drive a sale in the right direction. If the front line salesperson is not comfortable for any reason talking to the BDM that supports them (fear they will lose control of the sale, worried about the BDM performance in front their client, etc.), the sales support line can play a reassuring voice to make sure the BDM is appropriately involved. Figure 4 below outlines the basic structure of managed services sales support.

Figure 4

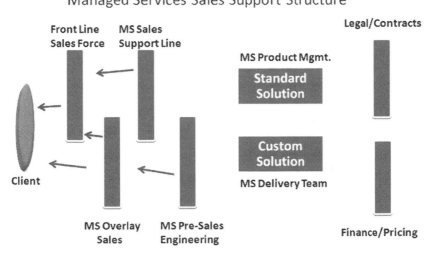

Managed Services Sales Support Structure

Finding the right people to man the above positions isn't easy. The BDM role is not pure sales; it is solution selling. You must have at least some technical proclivity. There must also be some overall business experience around contract negotiations as well. The importance to the client and your business on key contractual items such as termination clauses, SLA's, indemnification clauses, limit of liability, etc., requires the BDM be able to lead discussions of these items, even as they get support from the appropriate contracts and legal people in their business.

There is no quick answer to where you find these people, it's not easy. One thing I have always done wherever I have been is to look outside your company to complement your team. A little bit of looking on LinkedIn can fairly easily lead to finding managed services sales professionals at other companies. Grooming your own internal talent, along with some targeted outside hires, can work very well. I do like the idea of taking more junior people and teaming them with an experienced person and have had success doing that quite often. As long as they are fully linked for a year with the same objectives, I have seen that work very well. Finding and grooming MS BDM's who have selling, technical, and overall business/contracting ability is worth the investment in managing very closely.

> I will discuss the role of offshore managed services delivery later in the book, but if you have an offshore delivery team, say in India, having some pre-sales technical support located there I have found to be very effective as well. Not only can you find great technical talent there, along with the obvious benefit of being close to the delivery team, but when teamed

with onshore pre-sales engineering, can lead to 24 hour productivity working on key bids.

The managed services BDM is a difficult job, and self-managing your time is crucial. In managing MS BDMs, there are several rules I have found very helpful. In addition to obviously managing to achieve their bookings objective and closely monitoring near term projected closes, each BDM should have a funnel (aka pipeline) objective. This objective should be a part of their formal performance plan, and their performance on this should be a part of their annual rating and raise. What I have found to be the best way to measure this is to look for each BDM (and your business overall) to have a funnel of deals with 50% or greater odds to close that is three times the bookings objective. I don't care where the deal is in the sales cycle, just if the BDM sees a 50% or greater chance of closing. I have often lowered the 50% number to 40% odds to close, as many BDMs can get nervous about calling a deal 50/50 to close.

The BDM funnel must be managed closely. If a BDM likes to keep deals quiet until they are close to closing, it is not good for the business. You want the funnel to be accurate, especially deals that have any degree of customization. It is very important that there is the capacity and skills in delivery to support any custom deal. There must be inspection and push in getting that funnel 3x the annual booking objective, just like the obvious push you will have to identity committed deals to close on an ongoing basis.

There are many reasons why the funnel is so important, especially for the more successful BDM's. Managed Services deals have typically long sales cycles (6-9 months is very good, 9-12 can be more typical for larger, custom deals) and often can be quite time consuming to close. It is very easy to put all your time and energy on your near term closes and spend less time on demand generation and the health of your funnel. I have seen many a BDM close a big deal and say to their manager "now I have to work on my funnel." That is simply not acceptable. The rule I have used is a BDM should spend 10-25% of their time to work the funnel. I don't care how busy they are; they have to do this. I spent the first three years of my career as a Managed Services BDM for IBM, what I did and had to do back in the mid-80s is basically unchanged 30 years later. I sent my orders in by fax and listened to a Walkman on the train to work, so that has changed, but the job fundamentals are the same. I vividly remember struggling to insure I had spent time generating demand and building the funnel and had to self-correct myself many times.

My main activity to generate demand in the 80's was to activate and work with the front line IBM sales force. For any MS BDM selling through a front line sales force, the same holds true today. I used to call it "knocking on desks," going to an IBM branch office (especially on Monday mornings when everyone was there). Now with virtual offices it's a bit different, but the fundamentals are the same. Making sure the front line sales reps know what to look for in their customer base for a good MS prospect and pushing them to talk to their clients to identify opportunities is a key action. How you handle yourself as a BDM will create an impression across the branch. You will quickly get labeled someone to trust and take out to clients, or not. You need to be someone who will help you make your number as a sales rep and put money in your pocket. Just as a managed services deal that goes bad can hurt you across the branch (or broader), not having a positive working relationship with a front lines sales rep can hurt your efforts badly as well. It can be an unfair judgment, but the front line sales rep is like a client, they are the ultimate judge and vote with their actions.

From a metrics standpoint for BDM's, I like to keep it simple for business development. I like to measure bookings and funnel size. There are other metrics to measure in business development, such as average time to quote, the length of your sales cycle and your close ratio, but in terms of measuring your BDM's, I prefer to keep it simple and focus on bookings and funnel size. As you outline your financial plan for the year, how much you will ask each BDM to sell is a key determining factor. How that target should be set can vary greatly from organization to organization. If you are selling through a front line sales force, how large your typical deal size is, how much support the BDM has, are among the factors that need to be considered in determining how much each BDM should close. I would say on average each year, the total, annual contract value is typically in the $.5-2m range, total contract value being determined by the average contract length you typically see. A pure play MSP would be in the lower end of that range, a larger managed services business in the higher end of that range. If your business is more oriented towards outsourcing, your BDM's will have larger targets than that.

Pushing the BDM's to drive their funnel is a key management responsibility. As I just mentioned, I do like to insure the funnel of higher odds deals (40-50% or greater) for each BDM to be about 2.5-3 times the bookings objective. It's a red light when you are lower than that, as you most likely have too many eggs in too few client baskets. If your overall funnel across your business is less than 2.5-3 times your overall bookings objective, it is time for the offer and business development teams to place a high priority on planning for increased demand generation.

The Game Plan

I would say overall that selling managed services is more a science than an art. There are a number of key things that must be done and core knowledge you have to have. When selling managed services, you need to follow the script with the client - listen intently and adapt your sales strategy as you go, focusing on fixing core client problems and enabling them to take better advantage of opportunities and challenges they have in front of them. This is a solution sale, you cannot enter the client engagement confident you know the answer, the answer comes from engaging the client.

Front Lines Sales Force Enablement:

As I mentioned, it's typical that the managed services business development manager (BDM) covers a set of branch offices or geographic territory. The BDM must really sell to two sets of people, the front line sales force they support, as well as the client. It's critical that each BDM have a game plan to "activate" the sales team in their territory. There must be good support from a central offer/product management team, but the most important thing is the BDM having the right 'on the ground' interaction with the front line sales team.

It's important to clearly communicate to the front line sales force the roles and responsibilities they have, as well as the managed services team, in achieving the managed services bookings objective. Certainly gaining agreement with Sales leadership on this is a must. Fundamentally, you want the front line salesperson to be able to identify an opportunity and shepherd it through the sales process with the client. Of course, it's important they stay involved given their relationship with the client. Any more than that, the front line account team will typically not be effective. Once you outline the role of the MS BDM, it's usually not an issue in terms of "who does what" in the sale. The front line account team has ultimate authority on what gets proposed and how it gets proposed, and that must always be the attitude of the BDM. It doesn't pay for the MS BDM to overstep their bounds; it never pays off in the long run. You can disagree on the sales strategy and that can be discussed with the front line Sales management, but it must be done in a way that doesn't fray the long term relationships or give managed services a bad name within the company.

When you are selling through a front line sales force, the MS BDM must own "activating" that front line sales force. They need to be constantly communicating to the sales force, sales managers, branch managers, region managers, all of whom need to be "touched" on a regular basis. The executive that manages your BDM's must also play a very active role in working with the sales executives in the sales force. Certainly a lot of it comes down to the BDM and their day-to-day interaction with the field. They do need deliverables and material from a central MS product management team, but they also must feel ownership for the successful activation of the sales team they support. There needs to be a series of presentations done to get the front line sales person comfortable with the role they need to play. The main things that need to be communicated include:

- Why you should propose managed services - the offensive and defensive reasons and the ability to better achieve your quota and make more money
- Overview of managed services capabilities, value proposition, differentiation
- How to target clients
- Key questions to ask to identity/qualify
- How to overcome objections
- Managed services support behind the BDM

In addition to presentations, there should be several deliverables the front line sales team should have. It's very helpful for them to have a "managed services sales playbook" that reviews all aspects of identifying and driving managed series sales. In addition, creating one page "cheat sheets" can be very effective, covering the why, what, and how of selling your managed services. These should be in addition to your offer overview materials. This "cheat sheet" or "battle card", is very concise material that outlines in bullet form why you should sell managed services, who to target, what the offer is, the value proposition, core differentiators, etc. I see this as a must for any front line sales force and other supporting groups to have.

Of course the material you present must be highly specific to the managed services offers and value proposition you are selling, as well as the dynamics of the sales force you are selling to. Below are some high level points you can include in your presentation and "playbook" that may be helpful. It's not an exhaustive list, but hopefully highlights some key points in each category.

⇨ Why Sell Managed Services

    o Elevate client relationship from vendor to partner by operating their infrastructure

    o MS can be catalyst for incremental sales of hardware/software/maintenance/ transport/ professional services, etc. - two thirds of managed services clients buy additional services from the vendor who is providing the service.

    o Changes discussion to "value" vs. "price" as you solution sell vs. product sell

    o Increased visibility and control of clients infrastructure, which can lead to upsell opportunities

    o Significantly increase retention / keeping competition away. Managed Services retention rates are typically 90+%

    o If you don't propose managed services/outsourcing/ cloud, your competition will deposition you.

⇨ Target Clients - potential criteria to review

    o Enterprise, Mid-Market and/or Small

- o Geographically dispersed (vs. large concentration of users/sites)

- o Identify target industries

- o Minimum # of Users/sites (if applicable)

- o Global or Local only

- o Aged infrastructure, ready for transformation

⇨ Sample Questions to ask your client (and listen!)

- o How do you feel about the cost to run XX (the part of their infrastructure your offer addresses).

- o What are your top priorities for the business around XX (part of infrastructure your offer addresses)? Do you have any innovation plans?

- o What percentage of your staff's time is spent fire-fighting vs. driving key initiatives?

- o Do you have all the technical expertise you need around XX (part of infrastructure you are offering). Do you get all the management information you need?

- o Make sure you understand the current technology makeup of the clients infrastructure and ask questions to fill in what you don't know (older and more multi-vendor can often be better) o   What demands are your users/customers placing on you now?

o   Are you looking to buy in an opex or capex mode going forward?

⇨   Objection Handling/Responses

Objections:

o   Managed services sounds like it would give me less control over my infrastructure.

o   I have no added budget to do this

o   We aren't looking to cut jobs

o   I bet there are many hidden costs with your proposal

Responses:

o   Managed services enables IT to

  ▪   Spend less time on day-to-day issues vs. more strategic areas

  ▪   Leverage best practices and change management processes

  ▪   Save money on headcount

o   Managed Services allows you to control your infrastructure instead of your infrastructure controlling you. You will now have time in a managed services relationship to strategically plan and drive your infrastructure where you want and need it to go

o   Managed Services reduces risk by controlling changes in your environment, providing extensive information to help resolve issues and provides more time for IT to focus on other business areas

o   Managed Services in most cases is less expensive overall - a TCO (total cost of ownership) model shows what a true cost comparison is, avoid staffing turnover/burnout

o   No need for jobs to be cut if you don't want, you can redistribute IT resources to drive more critical and strategic initiatives and increase job satisfaction as more mundane work is done by us.

Compensation Plans:

Insuring the front lines sales force and the MS overlay sales team are on a compensation plan that is appropriate is clearly a must-do. I've had all kinds of plans in place for both over the years and have come to some consensus regarding what is optimal.

The front line sales compensation for selling managed services must be tangible enough to matter, but of course not detract from their primary mission. You must compensate on bookings vs. revenue generated in-year for managed services. The lead times for a sale and the time to implement and recognize revenue are simply too long to compensate on anything else. I prefer to compensate for total contract value (TCV) with a maximum of three years credit. A TCV measurement insures your team in incented to drive multi-year deals, however I think having a maximum of three years makes sense. With renewal rates that should be 90% or greater, a four year deal is simply not even close to twice as beneficial to the business as a two year deal. You can also compensate based on annual contract value (ACV), but you must insure you are doing one year deals and two year deals only by extreme exception. Depending on their overall sales objective, a full three years of credit may be viewed as too much, so getting one or two years of credit up front may be more appropriate. The front line sales force executive management will not want their reps to be able to achieve their target by selling too much managed services, so this needs to be negotiated. Overall, from a compensation standpoint for the front line sales force, there are three keys: 1) getting bookings (vs. revenue) credit, 2) having that booking credit count toward achieving their quota, and 3) having that bookings credit help them earn compensation just as their core products do.

The process and decision on how to compensate the front line sales force is a critical success factor for any managed services organization, without it being done correctly, it can be a non-starter. Support for managed services from the very top of the business must also be there, and how you compensate the front line sales force is a key litmus test to the support you have.

The same basics apply for the compensation plans for the MS overlay sales force (the BDM's). You must compensate on bookings, again I prefer TCV with a three year maximum, though ACV is fine as long as the right controls to insure you don't do many (any) one year contracts are in place. It's not easy to have compensation plans for managed services overlay sales teams, it typically leads to a huge variance between your top and lowest earners, and the sales cycle time can leave many doing great work, but not realizing that in their earnings. It's for this reason that I prefer an "80/20" or "70/30" comp plan for the MS salesperson. (80/20, 70/30 refers to the amount of base salary paid to the sales person, that's the 80 or 70, and the amount of salary that is "at risk" for sales compensation bonus, that's the 20 or 30). "60/40" plans can lead to too much disparity in the earnings that are not always reflective of the value to the business. Earning about 30-50% of your base salary as your compensation bonus is what I have targeted typically for the MS overlay sales team. New MS sales professionals should often be put on a six-month draw, depending on the near term opportunity of their territory. Each BDM should have an individual objective, ie., TCV closed, and they can be compensated on percent of attainment of that goal or a $X per $Y of TCV arrangement.

I feel the MS pre-sales engineering and sales support line teams should be on an 80/20 or 90/10 compensation plan based on the overall managed services booking goal of the business. If that is not possible, a bonus plan could be put in place where the team can earn 5-10% of their base salary based on the bookings performance of the business.

I have heard of some managed services organizations compensating their overlay MS sales teams with a percent of annuity revenue, where the BDM accumulates a building and growing commission over time as they add clients. This does require a higher initial base salary to make up for the time it takes to build the territory annuity revenue. I personally don't like this approach. I want the BDM's hunting and moving on. Not compensating on bookings means there is little a BDM can do to impact their performance in-year during the second half of the year. Making sure the clients they have sold already grow becomes more important to their pocketbook than selling a new client. But I can certainly see the motivation to making sure BDM's cultivate and stay with their sold clients to insure they are maximizing their use of the service.

I can't emphasize enough how important this whole area of compensation is for managed services sales. And yes, it truly is a good barometer of the support your company has for your managed services business.

The Sale

The selling of managed services is quite different than a technology or tele-communications sale, and the person talking to a potential MS client must clearly understand that. This is a solution sell, you are crafting a solution that hits a client's pain points vs. selling a product offering with certain features for a price.

A Compelling Event:

One of the best managed services sales executives I have worked with over the years, Charlie Mantione, always asked his BDM's "what is the compelling event that will cause this client to buy." I think that describes very well a key aspect of selling managed services. It is not about showing up at a client meeting and describing your offers and the value and benefits they bring. It's about meeting with the right level client, asking the right questions, listening very closely as you ask clarifying questions, and fully understanding their pain points and strategic imperatives - then and only then can you begin to think about how to communicate a potential solution you may have for them.

Managed services sales tend to be either managed services-attach sales or managed services-led sales. An "attach" sale is when you as a vendor are selling a technology or transport solution and "attaching" the sale of a managed service on top of it as part of one sale. A managed services led sale is one where the catalyst for the sale is the actual managed service. In an attach sale, the actual technology or transport solution is the compelling event. Most managed services sales are managed services-led sales, so the compelling event is key.

Finding that compelling event is what you are looking for: are they about to do a technology transformation, make a change in their WAN (wide area network) provider, having to make significant staff and/or budget cuts, about to have an M&A event, and /or dealing with a user revolt over the services they provide; these are all items that could be the catalyst to look to a managed solution. A full understanding of their environment is imperative up front. As the managed services provider, it's your job to be a solution to the key issue(s) they have. And in the process, you need to find that compelling event to ride on and sell your service. Often it's important to make sure your prospective clients know you and are aware of your services, so when that compelling event happens, they think of you as a potential solution.

The BDM in the process must find a way to add value in the sale. They need to become increasingly seen as an added-value resource to the client, someone who can come to the table with ideas to fix key problems and/or find ways for the client to better take advantage of opportunities they see in front of them.

Creating a Compelling Event:

A great way to think about generating demand for managed services is figuring out how you can create a compelling event for your prospective clients. One program I have had a lot of success with is to do a free trial. You can run a program where you will implement a client for free, for a specified standard set of services, and the client doesn't pay for it for three to six months. After that period they can cancel or pay the ongoing monthly fee. You have them sign a three year deal with your standard service description and include a clear and unambiguous 'out' in the contract with no strings attached and no payment if they opt out at the end of the trial period. I know that sounds risky and possibly reckless as the costs to implement any client are significant. However, what you do need to appreciate is how much work the client must do to implement the trial. It really is no less work on their part than what they have to do to implement a regular managed services contract. Their 'investment' in getting the trial going will help them justify at the end of the trial why they are going to stick with your service. And once they can see and appreciate the service, it can become very "sticky" and hard to cancel. The three times I have implemented a program like this I have had very few clients cancel, under 5% for sure. And each time we did it, we felt at the end of the program it was the right decision. It could be something you consider to create a catalytic event to get the client to move to your managed services solution.

The first time I was a part of doing this was at the IBM Information Network in the late eighties. Our managed services business was struggling and many of us in the unit were wondering if we could grow the business fast enough to keep IBM interested in keeping the business. As a young staffer at the time, I and a colleague of mine, Stu Bean, proposed a plan to Syd Heaton, the General Manager of the IBM Information Network and one of the true great pioneers of managed services, to give away a mainframe "host connection" to our network for free for six months for IBM clients. This meant that the clients' entire user population, wherever they were, had access to our connectivity and value-added services. It was a pretty big investment on IBM's part, as it was not only an investment in labor to set the client up, but in those days you needed a private line to connect to our network vs. today's VPN connections. The company decided to make the investment in the plan, and it was a huge success. We kept almost all the clients who signed up, and it was a key catalyst for the growth we experienced in the division in the late eighties and early nineties. That was a long time ago with a very different set of services and circumstances, but having applied similar logic in several other programs at different companies, I can say confidently that done correctly, a "try and buy"-like approach can work very well.

Cost Cost Cost:

There are many reasons why clients buy a managed service - to improve the performance of their infrastructure, reduce the risk in running their environment or in a transition they are doing, speed up a critical transformation, etc. - all can be a reason to buy a managed service. But just as it's location, location, location in real estate, its cost, cost, cost in managed services. That's not price, it's the client's overall cost, two very different things.

By far the number one reason a client buys a managed service is to reduce the total cost of ownership of running their infrastructure. Your price as a managed services vendor is only one piece of that. The total cost of running their environment includes their staffing costs, third party costs, tools cost, and transport costs, among many other potential items. It is imperative as a managed services vendor that you convince a potential client that your service can reduce their total cost of ownership.

I believe every managed services organization should create its own total cost of ownership (TCO) tool. This is a tool where you get the client to answer a set of questions about their environment and it calculates the costs they have with and without your solution. Yes, it can be an overly academic exercise. But the action of putting together this analysis can go a long way towards convincing a client of your solution. It also can be used by them to justify to others in their company of the decision to use your service.

The creation of your own TCO tool can really help crystallize for you as a vendor the key benefits of your service, or illuminate inadequacies in what you are providing. You must have firm evidence to first convince yourself as the provider that your solution will reduce a client's total cost of ownership. Of course, that starts with the features of the support you are providing, on the ITIL stack, which the client will no longer have to do. This reduction in client headcount cost can enable them to redirect headcount to other more strategic functions or make staffing cuts. And getting the client to acknowledge and recognize that is a must. Tools are also a key component. Many clients recognize they need to improve their underlying monitoring and management of their environment. There is a cost they must bear in having the right tools. There are often costs a client pays to various third parties that can be eliminated with your service, be it for monitoring certain elements of the infrastructure or providing support the client is not able to do. It is important, whether you use a TCO tool or not, to have these discussions with the client.

I do feel the managed services product management team should work a customized TCO tool for the MS sales support team. The TCO tool summary below outlines the basic elements of a TCO tool that can be built.

TCO Tool:

This next section outlines in some detail the parameters of how you can create a TCO tool that works for your business. Bill Strain, the CTO at AimNet, built a very effective tool we used there. I have used a similar model at most of the places I have been. Here I have outlined a model with two main parts: an output for clients that is relatively easy to read and understand in summary form, as well as the detailed assumptions and calculations (input) behind the output. The backup data and calculations are often very critical to review with the client and ensure they understand them fully, so that they will truly accept the analysis and feel comfortable they are saving money.

The tool can be based off a simple Excel spreadsheet, though it really isn't that easy to build. The more work and detail you put into it, the more effective the tool will be. You certainly want it to be customizable for any client to input their data, yet also be able to show a total cost comparison even if the client is not able or willing to share key data.

TCO Tool Summary:

Input Summary - The client will input the data outlined below.

New or Existing Infrastructure

New or Existing Management Tools (do they need to invest in new management tools)

Efficiency Rating (excellent, satisfactory, poor) - self graded by client (will come back to this later)

Number of Large/Campus Locations

Number of Remote Locations

Numbers of Users

Device Counts

Desktops

Servers

Routers

Switches

Voice PBX's/Servers

(You can have some default number of the above devices per X users if the client only has user counts and does not have their exact device count)

Vendor Managed Services Offer (ie., gold, silver, bronze)

Contract Length

Output Summary - This section summarizes the output, comparing the client's computed costs vs. the vendor (your) cost. The actual cost of your solution will be based on the volume information in the input section multiplied by your pricing. The client's cost is more complex to compute and is based on the assumptions and analytics outlined later in this section.

One Time Costs:	Client Managed	Vendor Managed
Mgmt. infrastructure set up		
Client Review and Legal Costs	n/a	
Design and Implementation Costs		

Total One Time Costs

Monthly Cost:	Client Managed	Vendor Managed

Client retained cost
(it's important to show client not all costs go away)

Vendor Cost

Client Vendor Cost

Total Monthly Cost

Total Cost Comparison

The above is the output of what should be a detailed set
of analytics underneath the tool's calculations. There are
several key areas which should be calculated based on a
set of assumptions, which I review below. As you build the
tool, it is a good idea to involve an outside consultant to help
with the numbers used in the analysis. That of course will
help credibility-wise as you discuss with the client how the
tool was developed. The following are key assumptions and
analytics used to calculate the above costs for client managed:

Client Labor Costs

	Number	Fully Loaded
Salary		

IT Manager

Level 3 Engineer

Level 2 Engineer

Level 1 Engineer

⇨ You should fill in the above salary information with
default numbers. The client can adjust your salary
assumptions as needed. Often you will get the client
to fill in the number of each, but you should have an
assumed number of Level 1-3 and management for a
small, medium or large client in case they don't have
the actual number of personnel they have in each area.
You can fill in the default number of people based on
the size of the infrastructure you are managing per the
below small, medium and large definitions.

Client Tools and Infrastructure Costs

This is used when the client needs to buy and install a new
management system(s) to run their infrastructure. You can
outline three support models here, depending on the client
size, and come up with a total cost for the client in each
scenario.

Small - 100 devices or less

Management Tools Cost

Ticketing Tool(s) Cost

Training Cost

Medium - 100-1000 devices

> Management Tool(s) Cost

> Ticketing Tool(s) Cost

> Training Cost

Large - Greater than 1000 devices

> Management Tool(s) Cost

> Ticketing Tool(s) Cost

> Training Cost

The above three scenarios include the one time and monthly costs for the tools plus training that you would need to calculate based on a set of assumptions and put into the tool. The large client (greater than 1000 devices) would also include some people assumptions to run the management platform. Again, this section only would apply to a client who is considering a new tool platform. It is important to note that many clients have bought management software and have never really implemented it, it's important to uncover that fact as you talk to your client. And you will find that happens quite often.

Recurring Management Cost Assumptions

This is really the heart of the assumptions and analytics section, where you make the most important financial assumptions on costs to the client when they run it themselves. The following can be used as an initial guide:

Infrastructure Cost - $Xk per year in space, etc., could use $10,000 as a default

Maintenance on Current Systems - Use 15-20% a year maintenance on tools cost, both for new and existing systems. If client can't supply the cost for their existing systems, you can use an assumed tools cost and take 15-20% of that.

Staffing (from the previous section, *the* key item).

> IT Manager (make assumption on amount per X devices). Example, use .5 HC (headcount) minimum, then additional .5 HC per 1000 devices under management

> Level 3 Engineer (make assumption on amount per devices). Example use 1 minimum up to 250 devices, then 1 for every additional 250 devices, increasing productivity as device count increases.

> Level 2 Engineer (make assumption on amount per devices). Example use 1 per 100 devices, increasing productivity as device count increases.

Level 1 Engineer/helps desk (make assumptions on amount per devices). Example use 1 per 50, increasing productivity as device count increases.

Consulting Budget - use an assumed amount of outside the consulting the client will need to help run their infrastructure, 10% of overall staffing costs can be the default.

IT Efficiency Rating

On an Excellent, Satisfactory, Poor scale, have the tool be built on satisfactory and have costs increase 20% if the efficiency is poor and decrease 20% if the rating is excellent. By efficiency, I mean a rating of how strong a client's tools, people and processes are in managing their infrastructure.

So you can see while this is a fairly detailed set of assumptions, it is a fairly easy model to create in Excel where the client can input as much as they can, and you will have defaults in place if they can't. This is a great way to fully engage the client on how and how much your solution can save. The number one reason clients move to a managed services model is a reduction in their total cost of ownership, so using this tool to get into this discussion I feel is a very worthwhile exercise, even if the client debates some of the assumptions. The process of discussing and debating the assumptions is typically a very productive one, and as you tweak the assumptions with the client, you can often find them helping to drive the decision to outsource. If they are going into it feeling they can't save money, it can be difficult, but you will certainly be no worse after doing this than your standing was before.

From a client standpoint, you will typically know, depending on the client executive you are dealing with, how much detail to review. For some, going right to the underlying assumptions to gain credibility in the analysis up front will be the way to go. And for others, getting to the high level output quickly to keep attention and interest in the tool will be the way to proceed.

As I mentioned, a key is to fully engage a product management team member(s) along with your delivery team in creating and validating the underlying assumptions. And as you train your sales team, they will acquire very good knowledge on how to talk to their client about the cost benefits of your solution, even if the client is not willing to do the actual TCO tool analysis. All in all, I feel this is a great exercise for your team to work on to insure your value proposition can "come alive" to the client using their information.

The Client/Vendor Interaction:

To me, a successful managed services sale occurs when the client and the vendor see this as a mutually beneficial transaction along two lines: cost and performance. The diagram below, figure 5, outlines what I am talking about. Many large clients do this evaluation in a fairly rigorous, documented way, and for some it's done more informally.

Of course, any sale starts with a relationship with the key decision maker and influencers. This is a highly strategic decision a company would be making, so having that trusted relationship is key. Ideally, as a vendor that has been there and been working with the decision maker on other parts of their IT infrastructure, you have a good position to start from.

Figure 5:

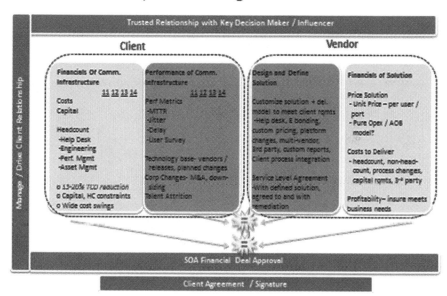

Anatomy of a Managed Services Sale

As mentioned earlier, cost is the key variable for the potential client. They have a view of the **financials** of the part of the infrastructure you are proposing your company manage. They also have a view of the capital they have spent and need to spend. The headcount they have supporting the infrastructure in various parts of their organization, whether dedicated or parts of headcount, is a key cost for them to consider as well. Of course, all companies are getting pressure to reduce IT costs and improve efficiencies, and I have found that typically in order to make the leap to managed services, most companies need to see a 15-20% reduction in their total cost of ownership. And they are looking to avoid what may be wide costs swings.

Other key considerations in the overall financial situation of the part of the IT infrastructure you are looking to manage may include any capital and or headcount constraints. One critical item to determine as a managed services vendor is how the client intends to procure technology going forward. The move of many companies to provide IT in an opex, or utility model (which we will discuss later), is gaining in popularity at a rapid rate. Many companies want to reduce their capital outlay and buy in a unit price that is inclusive of technology and support, what I have termed a utility model. I have found that most companies have already made this decision, they either want to buy in a traditional capital model or are moving to utility model, be it a public or private cloud environment. Headcount is another key consideration for the client, with many company's IT teams under headcount constraints. As user needs and business imperatives keep growing, and headcount remains flat or reducing, it can create a true inflection point for many companies where they need to do something different. A holistic view of financials of the different parts of their IT infrastructure needs to be evaluated by each potential MS client vs. the profile of the financials that would occur with your proposed solution.

The other dimension a potential client must fully consider is the **performance** of the part of the IT infrastructure you are discussing with them, shown in the darker shade in Figure 5. Your client has a view, whether they track it with documented metrics or more anecdotally, on availability, response time, and user satisfaction. They also have a view of the technology and WAN services base they have already and what changes and investments they need to make going forward. The age of the equipment, performance of their WAN services, how they do change management are all things they continually will be evaluating. The overall performance of the specific part of the client's infrastructure you are bidding on can be impeded significantly by downsizing, M&A activity and other broader corporate initiatives, creating an opportunity for a vendor. The potential MS client will be looking for improved performance of their infrastructure in moving to a managed vendor, and all these aspects are a part of what they need to consider.

From a vendor standpoint (the right half of Figure 5), it's your job to have the equation work along both the financial and performance dimensions.

The **financials** of the managed service need to meet the client's goals (ie., cost reduction, capital spend). The price you charge in addition to their other costs, such as headcount and third party costs, must lead to a total cost of ownership that meets the clients requirements, often a minimum of a 15-20% savings. In addition, if the client is looking for a utility price inclusive of technology, you must meet that requirement as well. The price you charge the client, whether it is your standard offer or a customized solution, must enable you to hit your profit objective. (We'll discuss pricing and profitability of managed services later). It's tempting to set a price that enables the client to hit their cost objective, but it cannot be one that will cause you to not meet your profit goals. When the price you charge hits the clients TCO objective, while at the same time meets your profit goals, you have made the first equation work.

The **performance** of the client's part of their IT infrastructure in question is that key second dimension, the second equation you must make work. You must convince the client that your offer will result in better performance. Generally, it is not too difficult to convince the client of this. They intuitively will believe that your scale, expertise, and tools will be superior to theirs, can they afford to buy your solution is typically "the" question they must answer for themselves. Having said that, there are still many things the client will want to inspect in how you will deliver the service. In general, the level of inspection will increase the larger and more custom the deal is and the level of sophistication of the client. Especially as deals get more custom, getting the client confident and comfortable with things such as your tools, reporting, engineering expertise and overall processes is very important, and can be challenging. They have goals they want to achieve in terms of performance metrics and user satisfaction, and they need to be confident you will enable them to achieve those. Often clients want documented service level agreements (SLA's) to give them peace of mind and make sure the vendor has skin in the game, which we will discuss later. As I mentioned, I am a strong believer in proactively proposing SLA's to clients with financial remediation. Clients generally want you to feel the pain when they do and want to make sure that you are fully motivated to hit key performance metrics. I have found that clients are typically fairly reasonable in negotiating SLA's, but as a vendor you should have a mini "center of excellence" on SLA management and know what types of SLA's you want to put in a deal and various terms that may over-expose your business.

As you outline the "hows" of providing your service and the client is convinced you can achieve their performance goals (and you are confident in your ability to deliver at the cost required), you have made the second equation work, as shown in Figure 5.

So there are two equations that must work to make the sale. The first is the financial equation and the second is the performance equation. Your BDM leading this sales effort must focus on adding value to the client decision maker - truly "adding value" by being someone that can make him/her more successful by focusing on fixing their problems and taking advantage of opportunities they may have. The BDM must have an insatiable desire to find ways to become a trusted advisor and a problem solver, which are required to solve the two critical equations. The result will be getting the order and beginning a multi-year journey with the client when the two equations work for both the client and you. This pre-sales evaluation process should give the client confidence in their decision.

By the way, it is a good idea, especially on larger deals, to have the client meet during the pre-sales phase key members of your delivery team who will be supporting them post close. Making these two equations work involves convincing the client and making sure you as the vendor fully understand your costs and ability to deliver the required solution. You must enter any client engagement with confidence in truly understanding those key underlying dimensions of your business. Your delivery team must be "all in" to your overall pre-sales efforts, and meeting the client provides motivation and incentive for the delivery team to feel they own or have a major stake in closing the deal and make the equations work! This is a topic will spend more time on later.

Another key area is the contract - how the contract should be structured and key negotiating items where you should draw the line. One thing that must be put in place is to have a legal/contracts person in your company focused on managed services contracts, someone who is available and comfortable negotiating with clients directly, which we will discuss more in the next section of this chapter

When to Walk:

One of the biggest skills in driving managed services sales is knowing when to walk away. It's easy to say you need to walk away when the signs from the client continue to move things to right, but it's very hard to do, especially as your 'investment' in the sale continues to grow over time. You need more than the BDM to help make this call. The BDM manager must play a role in evaluating when to walk, the BDM is often just too close to accurately make that call. There's no magic in determining when to walk, you just need to push yourself to evaluate it closely as weeks turn into months.

One dimension of evaluating when to walk away is when the buying signs and 'compelling event' are just not there from the client. There are several other things to consider as well.

One important thing you must have as a managed services vendor is a view that permeates your business on the need to figure out when walk away. That is the most critical first step, you must actively acknowledge and prioritize this as part of your sales management. I think it is safe to say that most managed businesses do a very poor job of this.

From a price and profit standpoint, you must have a pathway you believe in on each deal to get where you need to be. You must push your entire organization to have documented, agreed-to plans to hit your profit objective on each prospective deal. As pricing gets negotiated downward with a client, things can get tough. When you say to yourself, "we really need this deal, it's a top line needle mover, and we can figure out how we can make this work cost-wise as we get into the deal" is a sure fire recipe for future issues. And it also is an abdication of management responsibility. If you have a good handle on your operations, you should be able to come up with a plan to hit the costs you need to.

From a delivery standpoint, you must have full confidence you can deliver to meet the client expectations, along with the SLA requirements of the deal. As a delivery leader, you do not have the option of saying the reason we are having delivery issues at a particular client is the poor job that was done during the contract phase by the sales team. What you are saying when you say that is *you* did a poor job having a process to insure you and your team have approved every deal that has been proposed.

Shortcutting any of the above in the desire to close the deal will hurt you every time. There is no hiding in managed services, no separation of "close" and "install". And there are no excuses for having a bad deal from a financial or delivery standpoint, only action plans to make sure it doesn't happen again. A stage you don't want to be in.

Being able and confident in when to walk is directly linked to the confidence you have in how you run your operation. The more you understand your true costs, your delivery processes and tools (and their potential and limits), the more you will be able to confidently know when to walk. The less you do, the more likely you will make the sometimes fatal mistake of accepting a deal that brings short term gratification in the closing of a big deal, but long term pain in terms of cost and client satisfaction. Losing that one big client is twice as painful as the joy of winning that client initially.

The Contract

The Managed Services contract is a key area that needs to be well thought out. Managed Services contracts are typically three years, but can go to five to seven years especially if you are putting assets on your books. The managed services offer and business development teams should work together to have an optimal contracting process. You must have a standard master agreement of some sort between you and the client, and the actual managed services contract is usually an addendum to that and is typically based on the SOW (statement of work). You have a core SOW for your standard deals which outlines exactly what you will do as the vendor and what responsibilities the client has. Custom deals will of course have a custom statement of work, which should be based off of the standard SOW.

The master agreement, sometimes called a Master Services Agreement (MSA), should outline the key non-managed services parts of the agreement between the two companies. This would include how you will handle invoicing and payment, orders, confidential information, intellectual property rights, indemnification, limit of liability (one year of billing max should be standard), contract assignment, force majeure, and solicitation among other key items. Typically, the term and termination information would be in the managed services addendum or SOW. You should be able to structure a global deal where there is a single global contract for the client to sign and still do in-country billing. You will likely need to list pricing by country in the SOW, as well as appropriate foreign exchange (FX) currency information. Forcing a client to sign a contract in each country they want to do business in should be something you try to avoid for obvious reasons, but may be required.

The real core of the agreement, of course, is the managed services addendum or SOW.

Here are the key parts of this document:

⇨ Offer/Functions provided - This first section typically starts with a detailed overview of the service elements you are providing. Any custom services are also outlined in this section.

⇨ Onboarding - reviewing what will be done in implementing the client to your service. One key is outlining when the service is considered to be installed and "live" to start billing.

⇨ SLA's - documenting the SLA's. This will be your standard SLA's as a default, but most large deals will have a custom SLA, which we will discuss later.

⇨ Change Control - a very key part of the agreement, reviewing the process for all changes and what you and the client will do. Typically defining different types of

MACD's (moves/adds/changes/deletes) is included here (ie., simple vs. complex MACD activity).

⇨ Client Responsibilities - another very important part of the agreement. You must review their responsibilities in areas such as addressing, site/equipment information, client help desk requirements, maintenance agreements required, release management responsibilities, client point of contacts, among others.

⇨ Pricing - outlining your standard prices for standard deals and custom pricing for non-standard deals.

⇨ Termination - one of the most critical parts of the agreement, discussed below.

The SOW is such an important document. It is critical this get reviewed at an operational level with the client in addition to a contractual one. Scope creep is a reality in many managed deals, where the client expects a level of support over and above what's in your SOW. While well-meaning folks on your team will want to please the client, from a profit standpoint, scope creep can kill and must be avoided. The best way to avoid it is insuring full clarity with the client up front in the contract and in the operations guide you develop for them, which we will discuss later.

Client deals that include you as the vendor having the assets on your books need a special addendum to deal with items such as title, asset buy back, assets at end of contract, and appropriate hardware/software license information.

Pricing, of course, is a very important part of the SOW. I have found it very difficult to get one time charges in managed services contracts, it often flies in the face of what the client not wanting to pay an up-front charge. They are often looking for a flexible, unit-based arrangement that is a pure Opex (operating expense) vs. Capex (capital expense) model. And if you try to force a client to pay an up-front charge, they can typically find a competitor of yours that won't. It is because of the lack of an upfront price that the cancelation clause in the contract is so important. Managed Services deals from a cost standpoint are heavy up front and lighter later in the deal. [One reason retention is so important is that profit on renewals, even with some price negotiation downward, is typically greater than the original contract.] Getting a client up and running on your platform takes a lot of work, and once it is up and running smoothly, your costs go way down, a major reason managed services deals are typically back-end loaded from a profit perspective.

One way to improve the revenue up front in deals is to try to get a higher unit price for some of the larger units. If the client is wanting a total "price time quantity" model with no up-front charges, you can look to have different types of unit pricing. For example, if a client has four main locations and say fifty remote locations, you could charge more from a unit price standpoint for the four locations. It is likely those key sites will be onboarded early in the deal, helping year one profitability. If you have an assets on books model, there may be a few core data centers where you could have a higher unit price and achieve that same thing, more revenue early in the deal.

One year deals with no up-front cost are an absolute must to avoid. And two to three year deals with no up-front costs must include a cancellation charge. Your contract simply must recover your one time costs, including any assets. The cancellation charge I have used most is 50% of the remaining charges of a contract, so the cancellation charge decreases over time. Starting with a standard termination charge of 75% of the remaining charges of the contract is probably a good idea, and then in larger custom deals you can get negotiated down to 50%. When you have assets on books in your contract, there are separate clauses that must be inserted in the contract relative to asset buy-back and disposal. Depending on the size of the asset pool, you may need special termination fees for the asset part of the deal.

The sales cycle for a large, custom managed services / outsourcing contracts can often by a 6-12+ month process, and it can get painful at the end. The legal aspects of closing the deal can often take quite a bit of time. The sooner in the process you send the client a sample contract, the better.

When you get back a "red-lined" version of the contract it can be a very deflating thing for the team trying to sell the deal. As I mentioned, having a good legal/contracts person to work the deal can be crucial. The team working with the client negotiating the deal should be aware of the key contractual terms and what flexibility you have where. For example, you cannot have full termination for convenience without an appropriate up front charge, period. You must have an acceptable termination charge if there are not sufficient up front charges to cover your costs, which, as mentioned, there most often is not. And there should be a "cure period" for any contractual breach to be in effect regarding missed SLA's.

You must know your "limits" on the limit of liability clause. The sooner any issues on key contractual items come up the better, and the more you discuss them with the business executive at the client (vs. their procurement/legal teams) the better as well. There are places the client may want you to go contractually that you must be willing to walk from. For the most part, in the end, clients will respect the few areas you draw hard lines in, and it will end up in discussions over gray areas that require negotiation. I always prefer a "get in the room and don't let anyone out till we have agreement" approach. Even though you don't usually finalize a deal that easily, you do tend to make tangible steps forward each time you have a session like that.

Other Topics:

I often get asked how much should I, as a vendor, invest in the appearance of my NOC (network operations center). I would use this rule of thumb, make it presentable so if you were a potential client you would feel comfortable. Don't spend any more than that. I have spent hundreds of thousands of dollars on smoked glass that automatically goes from opaque to clear as you look in your NOC office. It's not worth it. When it breaks and needs to be fixed, it's doubly not worth it. I like having a large, well-equipped conference room with a view into your NOC (a nice curtain is fine). Clients want to be convinced you have invested well in your platform, processes and training, not in your glitzy presentations or overly fancy furniture and fixtures.

One item all members of your pre-sales team must internalize are the key differentiators of your service vs. your key competitors. Those key items should permeate everything you do and communicate. We discussed this in the offer section, but having your entire organization "singing from the same hymn book" on this, and constantly "proving" it in different pre-sales efforts is critical. The entire extended pre-sales team needs to be well-versed in this.

In Sum

Like almost all parts of managed services, selling managed services is more science than art, more substance and less flash. You must keep your eye on the fundamentals.

- All client facing managed services pre-sales resources must have an in-depth knowledge of the technology area and your offering.

- Have a relentless pursuit of asking open-ended questions and listening; it's how you can get your managed service to be viewed as an answer, not a question. Don't enter the conversation believing you have the exact right offering for the client, let it end up there.

- Make sure you have the right sales structure set up - a dedicated MS sales support team, a front line sales force incented and 'activated' to identify and drive MS opportunities and a strong technical pre-sales engineering team

- Have a well-documented offer to apply to key client needs, and a well-defined process to engage a motivated delivery team to work custom deals. The process and people must be in place to make this work.

- The true inflection point of the sale is when you get into the discussion and evaluation of reducing the client's total cost of ownership and performance improvement - when you are at the point of being viewed as an extended part of the client's team that can help them reduce their cost and improve performance, you are 90% of the way there. Getting to that point of helping them truly decide is when you truly become a trusted advisor.

- Do your homework up front to insure you have a solid contract and strong legal support. It is very beneficial to get a draft contract in front of the client as soon as you can in the deal cycle. Key members of your client engagement team should be well versed in the important parts of the contractual negotiation, including SLA's, pricing/billing and termination charges. Don't hesitate to make your key legal/ contractual points early in the sale with the decision maker.

- Proactively understand the parameters of when you need to walk away from a deal and stick to it. "Closing your eyes and hoping" the sexy big deal you want (and need) to close will work out from a cost and delivery standpoint will come back to bite you every time.

Chapter 5 - Service Delivery

Everything comes together as you deliver the service. Here is where you make or break your business, figuring out a model that works for you - delivering a level of support your clients need, at the required unit costs to enable you to be price competitive and make the profit you desire. And all the while, making sure you can scale the model and improve the efficiency and quality as your volume grows. If you've made it this far in the book, you clearly have a real interest in managed services. If you finish this section and have a desire to read on, you are a full-fledged managed services junkie.

The key topics we will cover in this chapter include **core operations** support, **platform**, **process**, **SLA's** (service level agreements) and **metrics**.

It doesn't actually take too much to deliver a basic managed service. Providing a differentiated level of support in a profitable fashion while growing strongly is very difficult. Most fail, or at least get stuck in terminal mediocrity, trying to convince themselves and others in their organization things are OK and no one could do a better job given whatever constraints they see themselves as having.

Overall, I'll start with this: delivering managed services is a high fixed cost/low variable cost business, and it needs to be run this way and be treated this way from the top of your business. Going into it with the attitude that as client deals are bid you will figure it out is a recipe for disaster. In the late nineties and early 2000's, as seemingly everyone looked to start-up an MSP business, the ranks of independent MSP's in the US grew from a small handful of companies to almost 100. After the economic downturn post 9/11, virtually all had failed and were out of business. There is no "winging it" in managed services, and many found that out the hard way. Less than ten survived by 2006. Since then, of course, the number of MSP's has swelled. Those that are successful have a well thought out and implemented plan *before* you have any clients.

While people cost is the highest cost you have in managed services, given the nature of the business, most headcount cost is truly not variable as in many other businesses. So while you may not think of headcount cost as "fixed," I have found in managed services much of it is. The high end technical talent to support all the technologies required, the 24/7 nature of the business and all the various functions you must have make a certain amount of your delivery headcount, in effect, fixed.

The biggest tangible investment is, of course, the management platform and tools. We will talk about the platform later, suffice it to say this is your core intellectual property that needs to be more than a few inexpensive off-the-shelf tools. In addition, your underlying processes around your offer definition must be well laid out. And your core NOC (network operations center) team, from your help desk through Level 3 engineering, must be in place for 24/7 support. Before you get your first client, there is a significant amount of cost and thought that must be invested in to make sure you can adequately meet client expectations and make money.

Core Operations

The core elements of delivering a managed service include having a platform that can remotely monitor and manage a client's infrastructure, a NOC where your delivery staff sits, and a set of processes documented to implement a client through all ongoing day two management. Let's review some of the core elements of delivering a managed service. I'm not going to go into too much detail on core NOC operations, rather focusing on key areas to make the managed service you offer a success.

Level 1 Support: You generally have a help desk or Level 1 support as your first point of contact with your clients. This team is typically dealing with a client's help desk or key support contacts vs. actual end users. I really think of this more as Level 1 support vs. a help desk. An end user help desk is a separate and distinct element from the Level 1 support I am referring to here. There should be very few calls coming into your NOC. 90-95%+ of all tickets opened in managed services support should be generated by your platform. Level 1 support will pick up the ticket to begin the process of resolution, communicating with the client along the way. There are several types of activities Level 1 will perform as a ticket gets opened. One, many tickets will simply be opening up a ticket with a WAN transport provider or hardware/software maintenance provider and then following up and tracking through resolution (all of which should have clearly documented processes driven by the platform). A second driver of activity which Level 1 needs to deal with is clients requesting a MACD. Your change management process with your clients should have clients contacting you electronically with a portal, or at least emails, vs. an actual phone call. And the process to fulfill that change should be highly automated. The platform should have the entitlements for each service clearly documented by client, so when a ticket is generated, the process to follow is clear, and you are not providing support the client hasn't contracted for. A third driver of activity for Level 1 is to resolve tickets which don't require technical investigation by Level 2 or 3 support. And a fourth driver is to notify clients of ticket completion, as well as updating clients on the status of more time consuming ticket resolution. The tickets that Level 1 can't resolve get moved up to Level 2 support. The types of tickets that Level 1 vs. 2 and 3 should resolve needs to be well documented and understood. It is good to document the percent of tickets you expect to be resolved by Level's 1, 2 and 3. A ratio of 70% resolved by Level 1, 20% by Level 2 and 10% by Level 3 is fairly typical.

Level 2/3 Engineering Support: There are many ways to organize the work of your higher level engineering team. It typically is beneficial to separate out your NOC engineering team above Level 1 into two levels of support, but it can also work with a single group of Level 2/3 engineers. How you organize what tickets get worked by whom is a key decision as well. It is typical to have your engineering team split by technical discipline, so for example your voice engineers and server engineers handle those tickets related to their area of expertise. But is often beneficial to organize along client lines, having a group of engineers who work on the same clients in a "designated" (vs. dedicated) fashion. This obviously enables a more intimate understanding of the client, but can be a bit more costly. You can also have a hybrid model, with some clients having a designated team supporting them and all others being handled by your engineering pool by technical discipline. There is really no rule to be applied to all situations, the specifics of your managed support need to be taken into consideration. If you have a broad range of technologies you are supporting, the more difficult it is to have a designated level of support. The larger and more complex your clients are, then there is more of a need to have a designated level of support.

Your Level 3 engineers truly play a critical role that needs careful attention. It is here where many of your higher value services are performed and a higher level of expertise is required. It is typically your Level 3 engineers who perform capacity and performance management functions, though sometimes you can have Level 2 do some of these tasks. However your core pool of engineers are organized, it's important for each of your most important clients to have a highly skilled engineer that knows their environment well - a "go to" person for the client when there is a key technical issue or question about their infrastructure. Having an assigned Level 3 engineer can become a key part of the client's experience. What I have seen is it is typical for a managed services client to highlight their assigned engineer as the most important part of their relationship with their managed services vendor. Which clients get an engineer assigned to them is an important call, as it is certainly a bit more costly, but I recommend that be costed-in as part of your service, at least for your most important clients. Having a strong set of talented, motivated Level 3 engineers should be a top priority of any managed services business.

Depending on the size of your operation, you may have a separate platform engineering team, or that role could be played by a few of your higher-end engineers. Given the overwhelming importance of the platform, it is highly preferred to have a separate team of platform architects. Often this team can become your "Level 4" engineers for the most highly critical or chronic client issues. This team can also play a big role in pre-sales situations, working with and helping convince key client technical contacts that you can meet their needs in a superior fashion to their other alternatives. But the platform team's primary function is to manage your internal systems, maintaining and tuning your platform applications.

Day 1 Team: The Day 1 team that implements and onboards your clients certainly plays a critical role in the overall managed services operation. *It's the first 90-180 days that will determine how successful and profitable the client experience will be for the life of that contract.* And that Day 1 team and their performance obviously plays a big part in that. The most basic element in Day 1 is the assignment of an overall project manager to run the implementation process. An "operations guide" (details below) should be created for every client, which outlines how the client will work with your operations team from a Day 1 and Day 2 standpoint. This includes escalation processes, notification processes/contacts, connectivity - basically outlining everything the client needs to know in working with your team. It is very important that all client data get collected and loaded in your platform 100% correctly. Many clients will not have a good handle on that information, and this must be dealt with aggressively up front. *Getting 100% accurate and complete client information for all sites and devices is a must, and the change control policies your employ must be air-tight.* Nothing bogs down Day 2 delivery or frustrates the client more than when problem resolution breaks down because of inaccurate data. The overall project manager must take full accountability for client data accuracy.

They also must make sure you have all the right resources assigned to the Day 1 team. If new technology is being deployed, the engineer responsible for the equipment design, as well as procurement, is a key part of the team. There may also be third parties involved in the actual new technology installation, which must be coordinated closely and effectively "turned up" and transitioned to Day 2 support. You may also have onboarding personnel who are responsible for entering all client data in the platform and insuring Day 2 management have all they need to effectively manage the client.

The person responsible for working with the client on connectivity to your platform and insuring all devices you are monitoring work with your tool-set can also be a key person, depending on the client situation. The security aspects of connectivity are becoming increasingly important to many clients, making your connectivity engineer, who is usually your security expert, a key member of your Day 1 team. If your Day 2 team is supporting a standard offer and the processes they will follow are fully documented and driven by the platform, the transition should be smooth. When the client contract includes custom elements, this must be fully documented and aligned with the platform. And Day 2 delivery management must formally assess the custom processes and documentation prior to the kickoff of Day 2 operations. There can often be custom development of your platform, and of course this must be fully vetted with the platform team. If there is custom billing involved, early involvement of the billing team is important of course, as this can be a painful item to deal with.

The project manager must do all the basics of effective project management - fully laying out the schedule with dates and owners, along with dependencies. And they must insure everyone gets regular communication on status. Some large implementations involving technology roll-outs can take years, so having an updated project plan that is documented between you and the client on what is to be done when and by whom is important to have at all times.

Given the importance of the first several months of Day 2 support, it can be very beneficial to have some key members of your Day 1 team transition into your Day 2 team, at least for a few months, to insure full knowledge transfer and help provide critical issue resolution.

Client Operations and Process Guide: As mentioned previously, it is important to have a document you and the client will use as your day by day agreement on how you are going to work together. *It is critical that expectations are set correctly between you and the client on all key operational parameters.* Prior to starting service, there should be a series of meetings and discussions with the client on the elements in this guide, filling out key information and making any agreed-to changes. There should be some form of client sign-off that they have read and understand the guide. This should be used by the delivery team as a "live" document and referred to regularly. Here are the key elements of this Operations and Process Guide:

- Level of Service Entitlement - This should outline for all technologies being supported what level of service the client will receive (and not receive, *very important!*). For each technology/offer, what you will do in monitoring, service desk, service management, incident management, problem management, availability management, configuration management and reporting needs to be outlined. It's not only what you will do, but what the client will do in each of these areas. This part of the document is a natural extension of the SOW in the contract, but should focus on key operational processes that will be followed. This part of the guide should be referred back to when the client looks for you to do more than you are contractually supposed to do, affectionately known as "scope creep."

- Escalation Contacts and Procedures - When there is a delivery issue, the process for escalating should be clearly documented, including who the key contacts are for the client. This could include the Level 3 engineer assigned to them, as well as first line delivery managers all the way up to the head of managed services delivery. There should also be an escalation path for overall account issues, which should include those responsible for overall account management.

- Letters of Authorization - When you as the managed services vendor will be acting on your client's behalf with other vendors that support their infrastructure, you need to insure that the appropriate letters of authorization are in place. This will enable you to legally act on your client's behalf in working with vendors, such as transport or maintenance providers.

- Service Level Agreements - Even though this is documented in the SOW, it should be outlined here as well for completeness. The specific reports and processes you will use to track with the client should be noted here as well.

- Root Cause Analysis (RCA) reports - For chronic issues (which can have a definition like three or more times over a one week period), as well as major incidents, you should proactively give the client an understanding of what you will do from a RCA perspective. Having this discussion and interlock up front can be very helpful when the inevitable issues crop up during day two operations.

- Notification Process - It should be clear to the client how and when and who you will be notifying when there is an issue. There are obviously many methods for notification that can be used (phone, email, client portal, text, etc), and insuring you have this interlocked with the client up front is important. As the vendor, pushing for electronic vs. verbal notification is obviously important from a resource standpoint.

- Change Management - Certainly this is one of the more important parts of the guide. The process must be fully documented here, including all approvals, testing, back-out plans, etc. There should be different types of changes (ie., simple, complex) documented, and what the client and you as the vendor are responsible for should be very clearly outlined.

- Onboarding Process - A checklist should be developed and documented on who does what for onboarding. The information required from the client on their infrastructure, as outlined above, is a critical deliverable.

- Portal - What information will be in the portal and how it should be used should be a part of this guide as well.

- Billing/Ordering - The process of 'what' and 'when' involving ordering and billing should be laid out clearly as well in this document.

Day 0 Team: In addition to your core NOC operations ("Day 2") and Day 1 teams, you need a small "Day 0" team to evaluate all standard offer deliverables and costing, along with custom client requirements. By Day 0, I am referring to "pre-sales." A small managed services operation could have these be roles played by key management personnel, but it doesn't take a very large business to require some dedicated resource aligned with this. The Day 0 team should be run by someone who is very familiar to your operation, someone who can do an effective job working with key leaders across your delivery team. This Day 0 team will typically work closely with your offer team as well, fully documenting and costing the different ITIL-levels of support for the various flavors of your standard offer. It will work with your business development and pre-sales engineering team on custom deals as well. It's in these custom deals that very quick, often very long-lasting, decisions will need to be made. It's important to have a delivery leader who is motivated to grow the business and not to keep the status-quo on delivery with minimal changes (and potential problems).

The Day 0 team will need to write, or validate closely, the core SOW deliverables, evaluating what functions are to be performed by your team vs. third parties. For activities you provide, the cost required to provide the level of support needed is of course a key deliverable of the Day 0 team. For many custom deals there can be tool enhancements required, so the platform team can be an important part of the equation. Custom billing, order entry, reporting are also often a part of larger bids, so the linkages the Day 0 team need to have can also extend outside of pure delivery. The Day 0 team will also play the critical role in finalizing the SLA's (service level agreements) with the client. We'll discuss SLA's in detail shortly, suffice it to say this responsibility is a very important one for the Day 0 delivery team.

The Platform

The graveyard for managed services businesses is quite large, and the cause of death more than any other is an inadequate platform. It's a slow but deadly killer. It also is truly what makes your offer come alive - it's how your team will support your clients, and it needs to be the core of the "intellectual property" and differentiation you bring to the market. In addition, it determines more than anything else how profitable you are.

Some of the largest managed services providers in the world have a boat-anchor as their core platform. Quite often they were originally developed ten plus years ago, and these platforms are the single biggest issue these managed businesses have. It hampers the ability to do custom deals and put out new offers. Often the costs are excessive to run and maintain, as well as update and upgrade. Yet as the business grows, it becomes simply unaffordable to replace and difficult to be competitive keeping it.

I say all this because if you are in a position to develop a new management platform for your business, you are very lucky, and you are about to take the biggest risk your managed business will ever take. A risk that most fail at, or end up in a terminally mediocre state. It's the single most important decision and project your managed business will undertake, and there is no quick and easy guide to follow to get it right. But there are many things you can do to increase your odds for success.

Certainly having a lead architect with experience in building a management platform is key. If you don't have one, stop, and find one the best one you can. Most managed services platforms at their foundation are built on off the shelf tools. The key to making your platform come alive is the "glue-ware", the coding and integration that goes across the piece parts of the platform to make it all come together. And it's your lead platform architect that needs to outline how the pieces will come together to be a high-functioning engine for your business.

There is no way in the space I have here to go over all that's needed to get this right. The technologies you support and the client base you are going after make the permutations quite large. But there are some core elements you need in any platform which I will review. At the most basic level, all platforms need the ability to do *remote monitoring* and have the intelligence to auto-generate tickets when problems are detected. 90-95% of all tickets your operations staff works on should be platform-generated vs. called in by the client. *Event correlation* is a must, so as to insure a single ticket gets generated when multiple parts of the infrastructure have an alarm due to a single issue or root cause. Without effective event correlation, a single issue can cause multiple tickets to get opened, and therefore worked, which means a lot of wasted effort of your delivery team. Overall, you want to fine tune the platform to insure it is not auto-generating too many non-client impacting tickets. It is important not to overwork your operations team needlessly from a ticket standpoint. Your *CMDB*, the configuration management data base, is a critical component. The accuracy of client data entered is so important here, as previously noted. Of course, you will need to insure the platform has the requisite *performance management* elements for the technologies you support. A layer of *predictive management* can be a real value-add, with tickets being auto generated as capacity and/or throughput thresholds are hit, as well as injecting synthetic transactions into the infrastructure and reporting out on potential issues. There are many management tools out there with all kinds of capabilities that should be researched and evaluated for use.

There are many other important parts of the platform. You need to insure you have an automated way to do *patch management* within the platform for most services. If you are managing desktops, being able to do *desktop policy management* for areas like power management and application blocking is key. Having a standard build for client desktops helps greatly here. If your service includes new desktop roll-outs, a remote imaging capability can be very helpful. Not only from a quality standpoint, but also minimizing the need for onsite client support during implementation.

A key piece of the platform is the *client portal*. Certainly you want clients to get comfortable using the portal vs. calling your staff, so making it as user friendly as possible is important. The money you spend on the portal will mostly go towards making it a pre-sales aid. The portal can be a real selling tool to help clients who are nervous about not having control or visibility into their infrastructure. The 'cool factor' can also be important as you are selling the client. However, the actual usage in production is always far less than what was envisioned by the client originally. The portal should nevertheless provide a clear window into the performance of applications, infrastructure and service levels. You certainly want the client to have access to a wide variety of reports from the portal. Your client management team, who are looking to insure the client understands the full value of the service you are providing, can leverage these reports to help the client recognize that value. If you sell your service through partners, the ability for those partners to brand the portal with their name can be very attractive. It is also important for a partner to be able to sell your portal as a key part of the value they will bring to their clients

At the other end of the sexy spectrum is the *billing and order management* part of the platform. Certainly, clients expect this to be there and typically don't inspect it. And this element of your platform can often have minimal investment up front. It is an important decision in the platform design, as the more manually you bill the more inefficient you will be. There are many very large managed services companies who regret not investing in enough in this component a generation ago, as retrofitting new billing platforms is about as painful as it gets. If your business is small and growing, it's important to keep an eye on billing before it gets too hard to change.

Other key elements of the platform include security, e-bonding and disaster recovery. A VPN tunnel is pretty much standard these days in terms of customer connectivity to the platform. And clients will typically want a review of your platform's underlying security measures. Clients and partners often require e-bonding from their platform to yours in more custom arrangements, so insuring you have standard e-bonding capabilities is required. If you are going to work with another provider in delivering your service as a private-label, e-bonding ticketing and inventory systems will certainly be a must.

One very expensive area is disaster recovery. Almost all MSP's say that have full disaster recovery in a dynamic or very short time interval, and most are pretty much lying. A fully dynamic, redundant disaster recovery arrangement is very costly, but having at least some sort of plan you can implement (and test) that can restore operations within a set period of time is a minimum.

A 2014 TSIA survey of managed services companies revealed that less than half of managed services providers have event correlation and a CMDB, while over 90% had a client portal. This means there is significant opportunity for many current managed services providers to improve the support they provide clients. The ability for vendors to differentiate with their platform is very high.

So what does it cost to build an effective platform? That of course is highly dependent on the breadth and depth of the services you are providing. I can tell you it's not a few hundred thousand dollars. On average, I would say the investment is $3-7m, though it can get quite a bit higher than that. Certainly you can work with vendors to spread out that cost so it's not completely an up-front investment. But yes, this is a big part of managed services being a high fixed cost, low variable cost business. Building a truly scalable platform that will be competitive for less than $3m is very tough. Hiring more "technical cowboys" to come in and save the day becomes the prime scaling lever in that case, and you are likely looking at a managed business that will run in the 25% gross margin or below range. When your business is in a steady state, the underlying cost of the platform should be around 5-15% of your total cost. 5% is very low, you have to be sure you have what you really need if you are at that level. Conversely 15% is high for a mature business. 10-12% is what I have found to be fairly typical. It's important to have a plan to get it to that level in what you envision to be a steady or mature state of your business.

One option you certainly have is to go to a more "public" platform provider, like a ServiceNow. These type managed platform providers can certainly do the job and often can cost less than building your own, depending on your requirements. However your ability to customize is low. And it will be very difficult to have a differentiated engine for your business. It is certainly a very viable option however.

I recognize this was a quick review of the platform, there is plenty of material out there to help put together an industry-leading management platform. I do think it begins with a trusted architect and a strong development team to build the glue-ware required. From a cost and finding programming skills quickly standpoint, it is good to have offshore development for much of the coding required. With the right architect, requirements definition, project management and documentation, the cost benefit of having this done offshore is worth it. There are several good offshore firms very capable of doing this work.

One of the most important things is insuring you have an approved business plan that includes the right level of investment in the platform. You can't get started without that, and your business will not succeed without it either.

Process

The basics of process documentation are a key part of running a successful managed services operation. All core processes from Level 1 through Level 3 engineering, inclusive of key functions such as change, configuration and performance management (among others), should be fully documented along ITIL lines. This is just another aspect of managed services that makes it a high fixed cost/low variable cost business, it's just something you have to do. As you outline and document your core delivery processes, those really become the foundation for your platform requirements. And in turn, it is your platform that must drive your processes. It sounds basic, but many managed services operations don't have their processes fully thought out or documented before building and adapting their platform. And similarly, many platforms that have been built end up determining the process flow of an operation by default, as you put the various parts of the platform together.

When I look at and evaluate a managed services business, the process documentation and the platform functionality, and how they work together, is a very key criteria.

Figure 6 below is an example of some very basic process documentation in terms of how a platform does alarm handling. It outlines how the managed services platform would flow for alarms at a high level. As mentioned, it is critical the platform drive your processes, and in turn your processes drive how your platform operates. Done correctly, the impact on the efficiency of your operations team will be significant.

Figure 6:

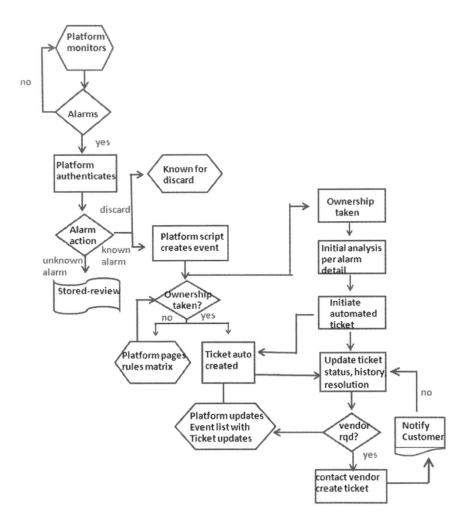

Managed Services, at its core, is not a cowboy business - by that I mean having really sharp engineers using their own personal expertise to save the day time and time again for clients. This is not a hero business. 80%+ of the people employed in your operation need to follow the process as outlined in your offer documentation and as driven by your platform (the other 20% doing high end engineering, planning, etc.). Doing that, with a great attitude, will give you the foundation of a great team. And you cannot overpay for this 80% of the work that's done - how much you pay for this support will go a long way towards determining your price competitiveness and profitability, more on that later . . .

You do need high-end, well paid technical folks. Those that architect the platform and the offer, and link those with your processes, will be some of your most important and well paid folks on the team. Certainly your high end engineers (aka level 3 / 4 engineers) are among the most valuable and well paid members of the team as well. As I mentioned, clients who get to know these high end engineers, and large clients should, will often identify those engineers as the most important value they get from your service.

Part of having complete process documentation is to measure both quality and efficiency. How many people (full time equivalents, FTE's) it takes to do the different items in the process needs to be tracked. In addition, you must identify key quality metrics. We will cover this in more detail shortly.

Service Level Agreements (SLA's)

Service Level Agreements have become an integral part of managed services offers. A set of standard SLA's with financial remediation should be a part of your standard offers. You can make sure your financial exposure is minimized by having your penalties be limited to, for example, the monthly billing of one site that you missed the SLA on for that month (vs. the entire bill).

As I've mentioned, I believe it makes sense to have an SLA "center of excellence" in your delivery team. There is an art and a science that comes with committing to penalties while not putting your business at undo risk. And if you do many larger, custom deals, SLA's are inevitably a key part of the SOW (statement of work). Negotiating the SLA can often become a very important, and lengthy, part of the negotiation. And that is why having someone who is adept at evaluating what you are willing to sign up for is so important, you need that core competency in your team to help close deals and limit your exposure.

The following are some examples of SLA structures.

Standard SLA's:

I have typically chosen 3-4 key metrics to track as part of a standard offer SLA. Examples of what can be used are:

- Notification - time it takes to notify the client (electronically typically) that a ticket has been created. Often minor tickets (low severity) are excluded from this calculation. Fifteen minutes is a typical objective for this, and achieving it ~95% of the time is about average.

-   Mean Time To Restore (MTTR) - the time it takes to fix an outage condition, typically the mean time to restore is in the 4-6 hour range.

-   Move/Add/Change/Delete (MACD) on time performance - An example of this is 24 hour turnaround, same day, for simple MACD's (if entered by a certain time) 95% of the time.

-   Availability - percentage of time the infrastructure being managed is up and running. 99.X% is typical on a site by site basis, as well as an overall metric for availability. The level of back-up and redundancy a client has needs be factored in of course to what you will commit to.

The delivery and offer teams should work closely together to come up with the SLA's and the financial penalties associated with missing the targets, as well as the prerequisites the client must have to qualify for SLA's with remediation. As I mentioned, in standard offer SLA's, care should be taken to limit any financial exposure for non-compliance. Limiting the penalty to a specific unit for a specific time period can be effective (ie., refund bill of site for a month of the site that drove the miss). You can also have the cumulative total for any misses to be a maximum of a certain percentage of the client's overall bill.

Custom SLA's:

Most larger clients will push for much more than you have in your standard SLA's, and this is where having an SLA center of excellence, even if it's just one person, can really pay off. Often the metrics being measured are the same ones as the standard offer, just the goals and penalties are more specific and severe.

As you do custom SLA's, it is often practical to distinguish between a critical outage and a less critical one. A critical outage can be defined as a total system down and major could be say 25% or more of the users and/or sites are down. It could also be tied to critical applications or systems being impacted.

As I mentioned, custom SLA's are often just having more aggressive targets, like critical outages MTTR being two vs. four hours. These days, I often have clients who come to the table with very elaborate SLA's with weights for various metrics being tracked (ie., putting more importance on certain metrics). As we get into these arrangements and negotiate, I will insist on an overall penalty cap (ie., 10% of one month's bill). Though more elaborate SLA's can be a pain, when you have a lot of variables it does give you some level of risk mitigation for you as the vendor. If you can negotiate a cap on each metric, it can really limit your exposure, though it does increase the complexity in tracking. Other metrics that can be asked to be tracked by clients in custom arrangements can include root cause analysis report timeliness, notification of utilization thresholds, integrity of asset information and timeliness of testing.

A successful negotiation of SLA's can really give the client a comfort level that they are protected and that you as the vendor will have true motivation after the contract is signed to perform well. I really like the idea of not shying away from SLA's, proactively going after it with the client and get an arrangement that works for both parties. I do find in the end clients tend to be more reasonable than not on this, especially if you get it to the right level of client. They just want you to be accountable and incented. If you can effectively limit your exposure during the negotiation, you should be in good shape.

Metrics

Determining your key operational metrics is an important decision. These must be metrics that measure both the quality and efficiency of your operation. I have seen so many managed services businesses drown in measuring too many quality metrics and be light on efficiency metrics. It's good to focus on measuring a small number of highly indicative quality metrics, as well as always including efficiency metrics, as part of your metric management.

Quality Metrics: There are some key metrics I have found the most useful, and most linked to client satisfaction. Percent proactive is one I swear by. That is the percent of tickets that are generated and opened by your platform vs. a client calling in. As a client, if you are using a managed service provider or outsourcer, it is extremely frustrating for you to identify an issue in your environment before your vendor does. I have found very little that frustrates a client more than that. Another key metric I've found very helpful is the percent of tickets that take greater than eight hours to resolve. Those incidents that are service impacting that take that long can be major dis-satisfiers. These are examples of metrics I prefer to focus on vs. other more common metrics that must be measured, but don't have the close correlation to client satisfaction in many cases. MTTR (mean time to repair/respond) is one which often lulls a business into sleep with good or improving metrics, but doesn't truly reflect the happiness of the client. Other items that must be tracked include notification time, MACD (move/add/change/delete) performance, on time implementation performance and availability. Of course these are metrics that are usually a part of your client SLA's.

Efficiency Metrics: It's important to place a priority on determining your efficiency metrics. This can slip to the bottom of the list, but it shouldn't. And you don't have to do complete time reporting by each of your employees to get there. At the most basic level, you should measure, for each technology offer, the amount of headcount it takes to manage a unit of that offer. So for your data network managed offer, how many FTE's (full time equivalents), does it take to manage X number of routers/switches. Similar metrics should be used for all your offers - number of FTE's per server managed, voice port managed, desktop, etc. The total number of FTE's in your metrics should add up to your total delivery headcount. It is very helpful to measure the trends in your efficiency by offer. When you use offshore support, I typically use a simple calculation that three offshore FTE's equal one onshore FTE. Headcount costs are not the only costs that impact efficiency, so measuring the non-headcount costs by offer unit is important as well and typically easier to track. Another worthwhile efficiency metric to track is cost per unit (site, port, whatever) to implement.

I have found asking the managers of each delivery department to estimate the amount of time their team spends on the different offers/technologies works well to determine the total headcount spend by offer/technology. They can look at key tickets worked, poll their people, etc. to get a close enough estimate of how their folks spend their time.

To this point we have focused mostly on Day 2 metrics, but you also must track Day 0 (pre-sale) and Day 1 (implementation) metrics. You should have reports on both areas done regularly (twice a month works well for these). For Day 0, documenting the key approvers of a large or custom deal, including design, implementation and Day 2 delivery, and where they are in the approval process by deal, is important. You need sign off from those three teams that the deal is ready to go. For standard offer deals, that should be a very basic checkmark, if any review is done at all. Each opportunity should include a review of the size and scope of the deal, so all parts of the business are aware. Tracking the cycle time for Day 0 deal approvals should be easy to report out on with this type of report being generated.

The Day 1 report should document the status of all key projects - who is assigned, the various key roles on the account and what the timeline is for the implementation (both original schedule and actuals). Any actions required by the client or your team should be detailed as well on this report. This document should enable you to insure you have the right resources in place to make the project successful. Tracking the cycle time and on time percentage for client implementations will come naturally from this report. The tricky thing, as we discussed earlier, is changing the schedule for client changes. It's important to keep the integrity of the on time metric by not altering the baseline to artificially inflate this metric. Your project teams can be tempted to blame a timeline moving to the right on client delays, when in fact it could be a breakdown in your support, so this must be watched carefully.

With these regular Day 0 and Day 1 reports, you will be able to effectively track your key deals and measure the appropriate metrics. We'll discuss metrics in greater detail later, especially related to how they all must be linked together across the various parts of the business.

Client Satisfaction Surveys: The question of client satisfaction surveys is an important one. Many managed services businesses are a part of larger companies who do annual, fairly elaborate surveys. I don't like these for managed services, never have and never will. The client gives feedback on all services the company provides, and it is often hard to distinguish what feedback is related to managed services vs. other parts of the business. Also, the clients who spend the significant amount of time required to do these long surveys are typically not the decision makers. My preference is for short, more regular "surveys" with the key decision makers. I like to ask for a simple letter grade on their satisfaction (could be basic numeric grade), plus a short commentary on strengths and weaknesses. And let them respond via email, voicemail, or whatever is the quickest and easiest way for them to respond. That's it, three to four times a year and you will have a great handle on the *true* client satisfaction of your account base.

To me, I've found the best goal to shoot for is a B+. If a client gives that grade, I think that is actually optimal. Many clients, even if they are very happy, want to keep you on your toes and insure that you not get complacent. B+ is the best you can probably get from them. If you get an A, you have to make sure you are not spending too much on that client. It's a fine line between having strong client satisfaction and making a profit. Regular client feedback is key to make you sure you are striking the right balance.

Managed Services is not something you can easily get great client satisfaction grades on. You are not developing some cool, highly needed application. You are generally not rolling out some leading-edge technology solution. This is plumbing - making sure your IT infrastructure works by supporting it well in a proactive and reactive fashion. One way to think about this is it's your job as a managed services vendor to monitor and fix things that break, and clients do not like it when things break. No matter how good you are, things break and you need to fix them. It's hard for some client's to feel truly delighted with this type of support. Clients who are not happy, giving a C+ or below grade, you should obsess about turning around if they are anything other than a very small client. If they are relatively happy, giving you a B range grade, you are where you want to be.

⇨ For some very large clients, no customer satisfaction survey can truly be enough to gauge how you are doing in the client's eyes. One thing several managed businesses I have been a part have done for very large clients is to have a "deep dive" analysis of your performance as a vendor and review it regularly with the client. You can only afford to do this for clients that truly are "needle moving" in your business, because it is terrible to start this type of review and not see it through each and every month or quarter. If you have a client that is one you must satisfy and is simply too big and important to lose, this type of exercise can pay big dividends. At IBM, we had a structured template we reviewed with all large outsourcing clients, and we developed the same thing at AT&T. More recently for me, we began an exercise like this for the top twenty Avaya managed clients which proved to be very effective. The managed services delivery team worked with the corporate quality team to develop a methodology which worked well, which we called a "drop cloth" review. Some corporate quality teams are better than others in an exercise like this, and the Avaya team was especially accretive to this exercise.

The work of grading how you are doing for large clients in key support areas, analyzing best and worst practices across accounts and communicating the scoring to your top clients can take your client relationships, and your ability to deliver, to another level.

Among the things you can analyze and grade yourself on, and then review with the client, include:

o   Design - strong pre-contract design review, any identified design issues in day 2, strong design plan for rest of contract, adequate testing of any design changes, etc.

o   Change Management - documented change plan, adequate client involvement pre-changes, adequate testing, etc.

o   Program Management - documented program plan, any date slips, future risk assessment, transition to day 2 delivery, etc.

o   Platform Monitoring and Management - any alarm misses, devices that can't be monitored, adequate redundancy to insure consistent monitoring, etc.

o   Client Leadership - regular executive reviews with the client, adequate support team assigned, internal financial and SOW monitoring and management, chronic issues not getting resolved, etc.

o   Service Level Performance - SLA achievement, adequate root cause analysis (RCA) for critical outages, etc.

A regular report like this cannot be undertaken lightly, as it is a lot of work that you must continue once you start. But coming up with your own scorecard for key clients, with your own grading system (including some of the key items outlined above), can be a unique, differentiated way to communicate with your client on the support you are providing (and the value they are getting!). As you can tell, focus of the evaluation is not only how good the support you are providing is, but also on how well prepared you are to avoid problems in the future. I think the idea of numerically grading your support along the dimensions above makes it more "real" to clients and employees and incents the right behavior and actions.

Other Key Delivery Topics

There are a few areas of managed services delivery that I feel are worth spending a few more minutes on, namely doing custom delivery, offshoring, onboarding, vendor management, delivery leaders and what I'll call avoiding the managed services "deadly circle".

Custom Delivery: While it is important to have a strong set of standard offers, you will invariably have deals that include support that is not a part of your standard offer. And, of course, these will typically be your biggest and best clients. Custom deals should only be done with large opportunities. Small custom deals kill. Resist the urge - tell your business development team up front you don't do small, custom. The only exception is when you are confident it will lead to a large custom deal. *Small custom deals kill managed services businesses!*

Large custom deals done right can ignite a managed services business. It can be dedicated support on or off-site, using a client's ticketing and/or inventory systems vs. yours, e-bonding, supporting non-standard technologies - it could be a myriad of things. It's important to keep to your core offer as much as possible and deviate off that foundation. The delivery team needs to be front and center in solution-ing and costing all custom requirements. It is also a very good idea to have the delivery team working the custom requirements meet directly with the client. In fact, that is often a prerequisite. As mentioned, it can also facilitate the sale, giving the client increasing confidence that you will be able to deliver.

Offshoring: In a word, yes. I know this is a controversial subject and many clients simply want onshore, "native" support. They also want a low price. To the point before that managed services is not a "cowboy" delivery operation, most people doing managed services delivery need to follow the process with a great attitude. The majority of those roles can be offshored to lower cost delivery centers as much as possible. Up to half to even two thirds of your delivery can typically be offshored. It doesn't have to be India, though that is the best place to start. Argentina, Central America, Philippines (for help desk), Eastern Europe are all viable options, each with their own strengths and weaknesses. Depending on where your centers are located, effective "follow the sun" management can take place with one center handing off to another as shifts change across the globe.

Done wrong and managed poorly, offshoring can be trouble. Done well, it can, and should, take your business to the next level profitability-wise.

It's important to have onshore-only options for those clients that demand it. And you can finesse how much a client actually needs to talk to you offshore personnel. But being able to have three or four skilled employees offshore at the cost of one onshore person is too powerful not to leverage. And it's not just Day 2 delivery. Offshore pre-sales technical and sales support, including pricing, can be extremely effective - especially in time sensitive pursuits where you can have virtually 24 hour support in turning around key deliverables.

You don't have to hire your own employees in India or other countries, you can work with many local vendors who can provide this support for you with a dedicated team. If you have been debating about doing offshore support, or doing it more than you are today, I would stop the debate and do it. To me one third of delivery should be offshored at a minimum.

Many managed services businesses focus on growing revenue with an assumption that new sales will drive efficiency in unit cost and strong profits. While volume helps greatly, it's your underlying efficiency that will minimize your unit costs and drive profit as you grow. What it costs you per unit to manage is just such a critical metric in my opinion. It's very difficult to be competitive in that metric without offshoring (and strong automation) being a part of the equation.

⇨ One question I get asked frequently, often by relatively mature managed businesses, is how to best start offshoring and how aggressive should you be in transitioning to it? Here is a quick primer on my thoughts. As you think about this, I would approach it from both a top-down and bottom-up perspective. What I mean by that is, it's important that the decision to leverage low cost labor be one that the leadership team of your managed business fully embraces. You must have a compelling story for your employees and clients on why you are going down this path, and you need to have a well thought out communications plan. The most important part, however, is the bottom-up planning that goes into what functions will be offshored. And that starts with your core delivery processes. You must examine closely each key process and identify parts of the process that you believe can be done effectively from a lower cost geography. In general, the more repeatable and process-driven a task is, the more it can be offshored. The more client interaction required, as well as the more interaction with other parts of your business, the less it can be offshored. Though certainly I don't mean to imply you shouldn't have offshore employees talk to clients or deal with areas that require internal communications. It's just that it will be easier to start with roles that have less of that. As you examine your core delivery processes, areas such as simple MAC's, configuration management, performance management, tier 1 / 2 support (driving / tracking ticket resolution, doing basic triage analysis, etc.) are among many areas you can easily offshore.

So how is it best to start your journey to offshoring? The first thing I would suggest is to have your offshore team be integrated into your various delivery teams vs. being separate and "doing their own thing." It's best to start with a small number of offshored folks be a part of each of your delivery teams. You could start with having 20-30% of your various teams (tier 1, tier 2, configuration, etc.) have offshore employees. They need to feel and be a part of each team, and your processes need to reflect that. As I mentioned, I don't like to have a separate team that is all offshored, say creating a simple MACD team that is 100% offshored. I think the percent that is on vs. off can vary greatly by department, but most departments in a managed services delivery team can leverage at least some offshore support.

One thing that cannot be overlooked is that having an on/offshore delivery model should increase client satisfaction, not hurt it. We talked earlier about the fact that problems that take a long time to solve, say severity one problems that take greater than eight hours to resolve, have a high impact on client sat. When each department has an offshore component to it, it's amazing how more productive you can be. That big issue your engineer is working in the afternoon can be taken over by a teammate offshore who is knowledgeable and ready to work it as their business day ends. You can get "true" twenty four hour support, which is especially beneficial on critical outages. Managed delivery operations that are all onshore have to admit their ability to fix key issues diminishes greatly after business hours. When a key onshore delivery person "wakes up" to find a critical issue fixed, the true value of having an offshore team will become very real to that employee, as well as the client, who wants an update first thing in the morning.

The best way to start an offshore component of your team is to get an agreement with an India-based IT services company, such as Cognizant, Wipro, Infosys, HCL, or Tata among many others. Clearly communicating the processes you want to be involved and skills you need is paramount. Possibly the most important thing is to get a team leader from one of those companies (and a back-up leader) that will run and manage the team. That is truly a critical role. I like having offshore employees report to a local manager and dotted line into their respective process owner, who is often onshore. So each delivery team of say 10-20 folks would have 3-6+ team members who are offshore, at least to start.

I believe a well thought out plan can be implemented in 90-120 days, and while there is some "bubble cost," you can begin to realize savings within a few months of going live with offshore support. It will save you money, and done correctly with good planning, improve (not reduce) your client satisfaction.

Onboarding: It is worth emphasizing the critical nature of your onboarding process. Simply put, the first three to six months post signature will determine the success of the client relationship for the duration of the contract - both from a satisfaction and profitability standpoint. Having a strong project manager run the process with the client is certainly very important. It is critical that all requisite client information is entered into your platform correctly and the client have a full understanding of the documented operations guide you will supply to them, which we discussed earlier. Expectation setting right up front is so important - how the escalation process works, how the flow of information and updates will work, etc.

A successful onboarding to Day 2, with you monitoring and managing the client's environment effectively, will result in a profitable relationship. You spend most of your cost up front in the client lifecycle, you need to get onboarding done and Day 2 running as quickly as possible to make the margins you need.

Vendor Management: Depending on how much of your cost base is outside to third parties, it may be worth a dedicated focus on managing the vendors assisting you in providing support. I'm a big believer in not "getting out ahead of your blocking" and using third parties to supplement your core skills. There is often no way you can have all the expertise required with your own employees. It provides a lot of flexibility from a cost and capability standpoint when you leverage third party suppliers. The managed services businesses I have run have had 20-50% of the cost base be outside third party providers, so its importance cannot be minimized. Insuring you are getting the best price and the quality you contracted for can often require highly focused support.

Delivery Leadership - One extremely important topic I have left to the end of this chapter may be the most important. The leadership of the delivery team is crucial - the person running the team must be fully motivated to grow and make money, as well as deliver a quality service. There is literally nothing that sucks the life out of a managed services business quite like when the delivery leader only wants to insure he/she has little issues, has as much headcount and cost as possible, and leaves any items related to growth and profitability to others. It is hard to find a delivery leader that truly wants to run a holistic delivery team, even when the going gets tough. Often the core qualities of a strong delivery leader who is motivated to have high client satisfaction and deliver iron-clad service don't extend to the overall business goals of growth and profit, but it most definitely is a key success factor.

The person doing this role is key, but also the actual role within the organization is important as well. In many larger organizations, the managed services delivery team can report to a broader delivery organization. That can work in the right set-up, but more often than not, over time, it leads to a delivery team focused mostly on delivery quality and getting as much headcount and funding as they can to do a good job delivering. That doesn't work in managed services. This is not a transactional sale; it is a solution sale. The solution formation for larger deals and critical offers must include the delivery team's active participation to work well, and it won't work well unless that team is motivated to drive revenue at accretive margins.

The leader of managed services delivery must be motivated and capable to drive additional revenue and margin, and they must be capable of driving results along those dimensions in addition to strong delivery performance. Who this leader reports to and how they are evaluated will, of course, directly impact that. The smaller the organization, the easier it is to make that all work, the larger the organization and company, it can get tough when decisions are made organizationally by some who really don't understand managed services and outsourcing. Nevertheless, it is a very important success factor for any managed services business.

Avoiding the Managed Services "Deadly Circle" - As I mentioned several times in this chapter, managed services is a high fixed cost, low variable cost business. You must have your offer, platform and processes well thought out up front - winging it simply doesn't work. The technologies you'll support, the functions you'll provide, the platform you'll leverage and the processes you will have your team follow all need to be outlined and interlocked up front. If you don't, and your theory is you will close a deal or two and adapt from there - using those deal(s) to learn more and help fund the investment required - you will find that is not a recipe for success. Once you get the client and begin the effort to support them, without a strong delivery foundation, you will likely find yourself losing money on the deal. In addition, you will most likely find that what's needed for the next client will have a number of differences that will require new elements to be completed to meet their needs.

In the build as you go philosophy, you will often find yourself "losing as you go" from a profit standpoint as the inevitable pressure will build to contain costs on the clients you support. As client issues come up, with the appropriate processes and platform support not being in place, your team will need to scramble to support them. The interest from the under-funded delivery team to get more new business will be low until the current issues are fixed. But what the delivery team will really need is volume to scale the business. However, without the right infrastructure from a people/platform/processes standpoint, the ability to do that will be limited or nonexistent. It is a deadly circle I have seen in smaller managed businesses all too often. Planning, preparing and investing in the scalability of your delivery team, platform and infrastructure up front is really the only way to avoid this deadly circle.

In Sum

Your offer and go to market efforts all come together in your delivery and client support, and there are a number of key items that need to be a part of the fabric of how you deliver your service.

- Managed Services overall, and especially delivery, should be approached as a high fixed cost, low variable cost business. Getting a few clients and then 'hoping' for efficient scaling without proper planning and investment is a recipe for terminal mediocrity and ultimately failure.

-   Your level 1-3 day two support team will spend more time with your clients than anyone, and while all of those roles are important, your level 3 senior engineers are truly the "backbone" of your support and are often the most valuable part of your service from a client standpoint.

-   The first 90-180 days will determine the success of the deal from a client and vendor standpoint, therefore your Day 1 team plays a critical role in supporting your clients. Especially important is insuring all data is entered in your system is accurate and the client is fully level set on what they can expect in terms of ongoing support, which should be clearly outlined in the Operations and Process Guide.

-   Delivery involvement in pre-sales is very important for all custom deals, including having an SLA center of excellence to insure you manage your SLA risk effectively.

-   Investing smartly in your platform and processes is a key success factor. Your processes, which should be fully documented, should drive your platform requirements, and, in turn, your platform should drive your processes.

-   As you manage your delivery effectiveness, make sure you monitor and manage your efficiency metrics in addition to quality metrics.

- The leader of your Delivery team, as well as the management underneath that person, must be motivated and measured on revenue and profit in addition to quality and cost management.

Chapter 6 - Client Management

Once a client is operational, an important decision that needs to be made is how to manage the overall relationship. It is costly to dedicate people to insure the client relationship is strong - namely that they are satisfied with the service and see the value. It can be tough in managed services for the client to appreciate the value without some ongoing communication. Having people dedicated to this function is something that should be "cost-ed" into your service. The overall governance model should be dictated to some degree by the size of the client. There are a number of "best practices" we will review in this chapter to insure client satisfaction, retention and upsell.

I see four major functions that need to be performed in Client Management.

1-Insuring the client is satisfied and focusing on driving any improvements needed, as well as finding solutions for any systemic issues.

2-Making sure your service level objectives are being met. Regular communication on how their infrastructure is performing is truly a must to insure clients are recognizing the value you provide.

3-Many clients expect their managed services provider to proactively make recommendations to improve their operating environment, this should be an ongoing focus of the Client Management team.

4-Managing the business relationship - insuring the client is on track to renew and "upselling" new managed services. There is definitely a "razor and blades" element to managed services clients. There is a tremendous ability to drive additional economic benefit as a vendor once you are managing a client's infrastructure. It can be in the form of new services provided, more volume consumed and/or chargeable change management.

A key question is how many people you should have in ongoing client management, not so much if you should staff this area. I've always thought of it in three buckets: large, medium and small accounts. Your medium sized accounts are where the decision on how to do client management is often so important.

I would define a large managed services account as one where the client is essentially outsourcing a part of their infrastructure to you. In this case, you are basically a part of the client's CIO team. The classic client management model here is the IBM "Project Executive" (PE) role, where a single person is responsible for the overall services relationship, including all aspects of client satisfaction and retention/upsell activity. This person is the single point of contact at an executive level for the client and the person who owns and manages the P&L for the managed services provider.

When selling to mid-market accounts, even though the revenue you realize may be small ($10-50k/year), for many of these clients you are in effect the client's IT team. Here you need this PE-like resource, but of course it can't be dedicated.

These smaller "outsourced" accounts really fit the "mid-size" client model I will describe below.

I am a big believer in assigning a PE-like senior delivery resource for large clients that fit the above description. Typically these large clients are extremely important to the overall managed services business. Having major issues can drag down the overall profitability of the account, in addition to client satisfaction. And, of course, losing a very large client can be very damaging to the business. So what constitutes a large client where the support is needed? In general, I think any client billing more than $500k a year should at least be evaluated whether this type of support is needed. Typically an account needs to bill well over $500k/year to be able to afford this type of PE-like support, but at $500k/year level is where at least some consideration should be given to a part time PE. The PE does not have to be dedicated, though I would not go to more than two to four accounts per PE. I have rarely had any PE have more than two accounts.

I don't really care for the title Project Executive, I prefer something like Client Support Executive (CSE). This role is one of the more important in the organization. Having your largest accounts satisfied and solid from a retention standpoint is the foundation of any high performing managed services business. Not to mention they are your best prospects for new business. I have found that it's really helpful to have someone who has been responsible for infrastructure delivery in some capacity in their past be the CSE. They must be very strong externally with the client and internally within the organization.

This CSE role must feel true overall ownership of the account, they cannot become adept at pointing fingers and informing the business of issues it has in supporting their client. It must play a leadership role in fixing any client-impacting issues. If there is a problem with the client and the CSE feels "it's not me they are concerned with," it's a problem. That role must feel complete ownership. The role must earn the respect of the client at senior levels. They need to become an extension of the CIO's team in insuring their part of the client's infrastructure is performing well and on budget, including regular reviews of the performance of what you are managing, inclusive of SLA results. And for any discussion about that part of the infrastructure you are managing, the CSE must be the "go to" person for the client. This role should be having regular meetings with the client, reviewing an agreed-to set of reports and making sure appropriate actions are taking place. For any areas of concern, a risk mitigation plan should be developed with the client and reviewed regularly. The role must *earn the right* with the client to be a *trusted advisor* and de-facto member of their team. And once that is accomplished, great things can happen from an upsell, reference and often public communications standpoint. Selling more to your large accounts should be a high priority of the business. According to TSIA, 60% of managed services contracts lead to an upsell of service and 85-90% of all future technology decisions are dictated by the managed services provider. There is typically nothing more important to a managed business than the relationship it has with its largest house accounts.

Your largest accounts which have a CSE or equivalent should not only have strong communications processes working with the client, you should also have strong communications processes internally reviewing how you are doing in supporting these clients. This internal review should not only assess how things are going, but also how prepared you are for future success. We discussed earlier an analysis of client support that can be a good way to self-examine and report on how you are doing. This can be a good complement to client satisfaction feedback you get either formally in a client sat feedback process or informally in client discussions.

Turning to the opposite side of the equation, let's review how to handle your smallest accounts. Obviously there is very little ongoing client management support that can be given to these accounts. They will all have standard service definitions, as small custom deals will crush any managed services business. To me, the key to ongoing management of these small clients happens when the day 2 service is turned up. The client needs to fully understand your operational processes and how they work, inclusive of change management and escalation processes. They should be comfortable using your client portal as well. I do feel it's important to have a small tele-support team for small accounts to do renewal management. This could be as small as one person, or a part of a person depending on the size of your business. This support should insure a client is notified well in advance (four to six months, or more) of when their contract is expiring to make sure they intend to renew. This will give you enough time to take actions required and work with the client should they be considering leaving.

In many ways, how you support your largest and smallest accounts is fairly obvious and intuitive. How you support your mid-size accounts is an important call each business needs to make. It's unaffordable to move to a PE or CSE-like model, and, by the same token, it is reckless to support them as a small, low touch client. Most importantly, proactively thinking about this set of accounts and how to best insure they are satisfied and will stay with you (and buy more) in an affordable model will most likely lead to the right going-forward answer.

What has worked for me in these mid-sized accounts is to have both a delivery and business interface. First, these clients do need a *delivery interface* they can call with any questions / concerns with the service. This role needs to also proactively communicate on how the infrastructure you are managing is performing. This can be done by assigning a senior, level 3 engineer to play this role. It can also be done by having a separate delivery manager assigned to the account. The delivery manager would typically have many accounts where they would be the client's interface. I've had a client delivery manager have 15-20 accounts, though 5-15 is ideal if affordable. Of course it all depends on the activity level required for each account. I prefer a delivery manager to a level 3 engineer play this role, but either can work.

There also needs to be a *business interface* for each of these mid-size accounts to insure retention and drive upsell. This client management role is basically a "farmer" role. Each person here can most likely also have 5-15 accounts, depending on client size and activity. Geographic proximity can also play a role in determining the level of responsibility for each person. You must determine how much support you feel is really needed to insure account retention, as well as how much opportunity there is to sell more to your existing mid-size accounts. There can be a real potential to have too little or too much resource on these accounts. So, as I said, the most important thing is to think about your support model from an effectiveness and affordability standpoint - as getting to the right model for your business is certainly a key success factor.

One thing worth considering is having a bonus plan for people outside your business development team for financial impacting items. And the people responsible for client management can fit in that category. Project Managers who are responsible for timely and efficient implementations are another, since they can accelerate revenue with strong performance. A bonus plan may be able to be developed within the person's normal bonus plan, but at many companies that is not possible. So assigning small bonuses, in the range of $1-5k, for things such has client upsell, hitting revenue retention objectives, hitting profitability targets or on time implementation targets can have a real positive impact. No, that's not a lot of money, but as you measure and track these types of results, it's amazing what a little recognition and money can do to impact the top and bottom line.

The "Stickiness" Factor:

We've discussed the high client retention associated with managed services contracts, and we will show how that plays into the financials of a managed services business later. Suffice it to say that one of the true staples of a strong managed services business is the "stickiness" of its client relationships. When you take over the day to day operations of a part of a client's IT infrastructure, you truly become a part of the client's team. And over time, the dependence on you as a vendor will grow and many opportunities should and will arise. It is one of the major benefits for any IT vendor of having a managed services business. And with the move to cloud and opex models, its importance is only growing.

Everything you do as a vendor, from the contract through delivery through ongoing client management, should be oriented around insuring clients will be motivated to stay with you. In many ways, it's a bit of a culture that needs to be fostered in all parts of a managed business. With a strong foundation of client retention and upsell, you are de-risking your managed business significantly and positioning it for future success as you continue sell new clients and deliver to their expectations.

Metrics:

Outside of core client satisfaction metrics, which we covered previously, it's important to track a few other key metrics around client management. One is certainly around retention. I like to use revenue retention as a percentage. As you enter any year with a certain run rate of business, what percent of that will you see in that year is the baseline of the metric. So if your business is exiting the previous year with an annual run rate of $20m from the set of accounts you have at the time, and from those accounts the following year they generate a combined $19m of revenue, you have a 95% revenue retention rate. I prefer this to a renewal calculation based on number of customers. Revenue retention makes the retention of a $20k/month client four times as important as retaining a $5k/month client. And what revenue retention should you have? The goal in my opinion should be 90%+. If you are in the 94-96% range you are doing very well. Low 90's is OK, but it must be watched closely, and below 90% is a big issue. Managed Services contracts are typically two to three years, but should last a generation.

You should also target an upsell objective each year. The standard goal I have always used is 3-5% uplift on your base set of accounts, 2% at the very least. One large upsell can blow away that objective, and conversely a few very large accounts with no upsell potential can depress your results. But upselling is a key initiative and opportunity and must be tracked and managed appropriately.

Chapter 7 - Strategic Operating Plan

Putting all the pieces together to have a managed services business running on all cylinders is a difficult challenge that requires very tight management. As I said, starting a managed services business and growing top line well (~10-15%+) is relatively speaking not an overwhelming task. Likewise, tightening up an already running managed services business to drive strong margins (30%+ gross, 15%+ EBITDA) is not a huge challenge either. However, having a growing managed services business that is generating strong profits in a highly scalable fashion is one of the most difficult things to do in IT. The next few chapters outlines what has helped me steer a managed services business ship effectively. There are several things I have found helpful in planning and running a managed services business: a detailed three year operating plan, an effective operational management discipline, a set of reports to track progress and a vehicle to inspect the business to evaluate the underlying effectiveness of the various parts of your business beneath the financial results.

I do believe effectively running a managed business starts with the three year strategic operating plan. We reviewed in the Offer section the overall strategic plan for the business - looking closely at the market (client requirements, competition, etc.), your offer, go-to-market plans, delivery, tools and client management to drive a competitively differentiated and sustainable value proposition and business. A key part of that strategic plan needs to be an accompanying three year operating plan. This can be viewed as a bit of an academic exercise, since three years is an eternity, but I have found it critical to have this as your guide to keep measuring your business against.

I use the offers defined (along ITIL lines) as the foundation for the strategic operating plan - dividing up your offer by technology and functions provided. As you closely evaluate your strategic operating plan, it's important to push yourself to specifics: how much of what you offer do you believe you can sell, what are the underlying costs and investments required each year, and coming up with a profit and cash flow plan that meets your business's goals. One key is "unit-izing" your business and continually looking to evaluate how you are doing vs. the plan (and competition) and adjusting as needed.

Measuring and managing the top line by technology or offer area is relatively easy. Planning the number and size of deals you will close, dividing that revenue stream by technology offer in a unitized fashion, gives you a good feel for what you need to drive through your factory. Closely monitoring your costs is tougher, but extremely important. The delivery team must analyze how much headcount and non-headcount costs they need to deliver that revenue stream, broken out by offer/technology and in turn into units. Much easier said than done, but an important exercise that I feel is required to drive an effective operational discipline into a managed services business.

A very important element of cost planning is determining the headcount that is required to run the business, given the volume growth and mix. One of the most important underlying metrics in any managed services business is the units managed per FTE (full time equivalent) headcount for each offer/technology area. [With offshore headcount costing much less, for the units managed per FTE metric, I typically count one headcount as three offshore heads.] This metric really pushes you to identify your underlying efficiency and scalability. It is much easier to determine your non-headcount costs in a unitized fashion - be it for basic maintenance support from third party companies, or specific engineering / tools needed to support a certain technology, or other support that you don't have within the business.

Sample Three Year Operating Plan:

The best way to make real what I'm talking about is probably to look at an example of how to put together a three year operational plan. I've put together a proforma three year plan for a managed services business, starting on page 138, with detailed assumptions for revenue, cost, expense, productivity, etc. By drilling down on an example, some of the key planning areas will better come to light. Regardless of the size of a managed services business, the key principles will apply.

As you can see in Table 1, this managed services business is planning to do $75m in year one revenue. It is currently growing at about 10%. The top line goal of this business is to grow at 15% and become a $100m business in three years organically. The business is running at a 30% gross margin and just under 10% EBITDA. The goal that's been set is to have a 35% gross margin and 15% EBITDA by year 3. It will take tight planning and strong execution to achieve these growth and profit goals.

This business has three major offers/technology areas - data, voice and server management. It's bread and butter has been data management, which has fueled much of its growth. However the expectation is there will be increasing commoditization (price pressure) and some reduction in the growth rate. The data business is profitable at 31% gross margin, and there is a strong feeling there is great opportunity to drive additional margin and move more work offshore.

The voice business in this example has been a tag-along business to the data, growing at a modest 5%. It's been a lower function offer, more monitoring and break-fix, with a relatively higher percent of cost to third party suppliers providing more maintenance-like support. But the business is seeing a trend of many clients having older legacy voice infrastructures which are ripe for transformation and a managed service. The plan this business has is to sell a higher value voice offer with more high-end managed services (performance management, enhanced multi-vendor support, etc.). The belief is this part of the business can be a growth engine, and it will require an investment in higher end engineering skills.

The server managed offer is a monitoring-like offering, with relatively low functionality. It has a high unit volume and is growing at 10% a year. The belief is this service line can maintain its growth and profitability, with some increase over time in both. Note that this offer here is a very low-level of service, where the client would be notified on key items such as when pre-identified thresholds on CPU and disk utilizations are hit. A more comprehensive managed offer including full incident management, problem resolution and remediation could cost hundreds of dollars a month and require much more headcount (FTE's) per server to support. The offer here only offers a very basic set of functionality.

SG&A Planning - You can see in the three year financial plan in Table 1 that this business will spend $16m in Year 1 on SG&A. It has a business development team it believes can generate $20m ACV/$60m TCV in bookings in year 1. I have found the following key functions are typically a part of a managed services SG&A, along with roughly how large.

Function	% of SG&A
Business Development (overlay Sales Force)	20-35%
Technical Pre-Sales Support	10-20%
Offer Management	15-20%
Platform/Tool Support	10-20%
Operations (quote to cash, sales ops, etc.)	5-10%
Executive	5-8%
Non-headcount (expense only)	10-20%

I have this business with 20 business development managers (BDM's), with an average of $1m ACV/$3m TCV being booked by each one annually. I have found on average about $1-2m ACV per BDM to be about right. But you will have huge variances by BDM, with a big deal or two on one side, and a virgin territory on the other, swinging results dramatically. The plan is to have the average deal size increasing from $1m ACV per BDM to around $1.5m per BDM in year 3.

Average deal size is a key factor in business development planning. This business has an average deal size of $150k

ACV, which means in year one an average BDM will need to close 6-7 deals. With sales cycles often at six months plus, you must have a pretty good clip of deals being worked.

Revenue Planning - The bottom of Table 1 outlines the build-up of the revenue plan by year. You start with the annualized run rate of the previous year. I typically use the final quarter of revenue x 4. You can see in year 1, in this example, that is expected to be $70m. Next is the assumed retention of your existing business. This business has been experiencing about 7% erosion year over year, and this is the assumed rate over the planning horizon. It's a very difficult, but obviously critical, thing to predict. 7% erosion is about average for a managed services business. That gives you your base revenue for the year, in this case $65m in year 1. Next you factor in the growth you expect from your existing accounts, in this example, it's $2m in year one. And last is the assumed revenue you will get from bookings you achieve in that year. I have found that on average you get about 5/12's of in-year revenue from your ACV (an average of five months of billing from your in-year bookings). In this case, with $20m of ACV bookings, you will see an estimated $8m of in-year revenue. All of this is shown in Table 1.

Cost Planning - From a cost and delivery standpoint, you can see a major focus is required over this three year horizon on improved productivity. This productivity needs to come from increased volume, improved automation (leveraging your platform) and an increased focus on moving work to low-cost geographies. This business has 50% of its headcount in low cost geographies today and needs to move that to almost 60% by year 3.

Table 1

3 Year Financial Plan ($m)

		Year 1	**Year 2**	**Year 3**
Revenue (% growth)		75.0m (10%)	86.0m (15%)	100.0m (15%)
Cost		52.0	58.0	65.0
Headcount		34.5	37.5	42.0
Non-Headcount		17.5	0.5	223.0
Gross Margin (%)		23.0 (30%)	28.0 (32%)	35.0 (35%)
SG & A		16.0	18.0	20.0
EBITDA (%)		7.0 (9%)	10.0 (11%)	15.0 (15%)

Headcount -	Total	428	501	565
	HCG	214	223	240
	LCG (%)	214 (50%)	268 (54%)	325 (58%)
Bookings		20.0ACV/60.0TCV	27.0ACV/81.0TCV	30.0ACV/ 90.0TCV
In Year Revenue*		8.0	11.0	12.0
Run Rate (going into Year)		70.0	78.0	90.0
Revenue Retention		93%	93%	93%
Base Revenue		65.0	72.5	84.0
Known Account Growth		2.0	2.5	4.0
New Rev. (sold/ billed in yr)		8.0	11.0	12.0
Total Revenue		75.0	86.0	100.0

*Assume 5/12 ACV results in revenue in year

Table 2

Year 1 Offer Financial Breakout

	Data	**Voice**	**Server**
Revenue	35.0m	20.0m	20.0m
# Units	29,000 routers	208,000 ports	33,000 servers
Price/Unit/Month	$100/router/mo	$8/port/mo	$50/server/mo
Cost	24.0m	15.0m	13.0m
Headcount	17.0m	9.0m	8.5m
Non-Hdct (% Cost)	7.0m(29%)	6.0m(40%)	4.5m(35%)
Gross Margin (%)	11.0m (31%)	5.0m (25%)	7.0m (35%)
Headcount - Total	210	112	106
HCG	105	56	53
LCG	105	56	53
Units Managed/FTE **	207/router/FTE	456 servers/FTE	2700 ports/FTE

HCG = high cost geography

LCG = low cost geography

*Assuming ~ $120K/yr HCG, $40k/yr LCG - any full time contractors are assumed to be headcount vs. non-headcount cost

** LCG = 1/3 of HCG headcount in productivity metrics

Table 3

3 Year Offer Plan ($m)

		Year 1	Year 2	Year 3
Total	Revenue	75.0	86.0	100.0
	Cost	52.0	58.0	65.0
	Headcount	34.5	37.5	42.0
	Non-Hdct (% of Cost)	17.5 (33%)	20.5 (35%)	23.0 (35%)
Data	Revenue (% growth)	35.0 (16%)	40.0 (14%)	45.0 (13%)
	Price/Unit/Mo (router)	100	97	95
	Cost	24.0	26.0	26.5
	Headcount	17.0	17.5	17.5
	Non-Headcnt (% Cost)	7.0 (29%)	8.5 (33%)	9.0 (34%)
	Gross Margin (%)	11.0 (31%)	14.0 (35%)	18.5 (41%)
	Units Managed (routers)	29,000	34,400	37,500
	Units Managed / FTE	207	237	272
	Headcount Total	210	235	245
	HCG	105	100	95
	LCG (%)	105 (50%)	135 (57%)	150 (61%)
Voice	Revenue (% growth)	20.0 (5%)	24.0 (20%)	30.0 (25%)
	Price/Unit/Mo (port)	8.0	8.25	8.50
	Cost	15.0	18.0	22.5
	Headcount	9.0	11.0	14.5
	Non-Headcnt (% Cost)	6.0 (40%)	7.0 (39%)	8.0 (36%)
	Gross Margin (%)	5.0 (25%)	6.0 (25%)	7.5 (25%)
	Units Managed (Ports)	208,000	242,000	294,000
	Units Managed/FTE	2,700	2,640	2,450
	Headcount Total	112	135	180
	HCG	56	70	90
	LCG(%)	56 (50%)	65 (48%)	90 (50%)

Table 3 (cont'd)

<u>3 Year Offer Plan ($m)</u>

		Year 1	Year 2	Year 3
Server	Revenue (% growth)	20.0 (10%)	22.0 (10%)	25.0 (14%)
	Price/Unit/Mo (server)	50.0	50.0	50.0
	Cost	13.0	14.0	16.0
	Headcount	8.5	9.0	10.0
	Non-Hdct (% cost)	4.5 (35%)	5.0 (35%)	6.0 (37%)
	Gross Margin ($)	7.0 (35%)	8.0 (36%)	9.0 (36%)
	Units Managed (servers)	33,000	36,666	41,666
	Units Managed / FTE	465	485	502
	Headcount Total	106	121	140
	HCG	53	53	55
	LCG (%)	53 (50%)	68 (56%)	85 (61%)

Table 2 gives a baseline view of the business by offer. You can see the relative sizes of the data, voice and server offers, along with the number of units under management in each area. There is also an average price per unit. This will obviously vary by the functionality provided. You can break out your offers into a gold/silver/bronze set of planning assumptions. I am using a single assumed average unit price for simplicity. You can also see the breakout of headcount and non-headcount cost by offer. For simplicity sake, I have assumed all offers have 50% of their headcount in a low cost geography in year one. From a cost standpoint, for this example, I am assuming each delivery headcount is $120K/year in high cost and $40K/year in low cost (fully loaded). At the bottom of the chart you can see the important units managed per FTE metric by offer. In calculating this metric, I am assuming three low cost heads equals one high cost head due to the cost implications. The baseline of units managed per FTE is a key metric to watch as you look to achieve your profitability goals, so getting the baseline right is important.

Table 3 gives a detailed view of the three year plan, with a focus on underlying delivery and cost metrics. You can see several significant things happening in this business. There is a huge focus on improving profitability in the data business. A major shift to offshore support is assumed, resulting in a significant increase in low cost support with, relatively speaking, a modest increase in headcount cost. A 10% improvement in gross margin is significant, and it requires a 30% productivity increase (207 routers/FTE in year 1 to 272 in year 3). You can see there is less leverage in non-headcount spending with scale, and the percentage of non-headcount cost grows over time. The ten point gross margin improvement comes despite an assumed reduction in average unit price due to increased competition and commoditization. So you have a lot going on in the data business. It's still growing with the pricing pressures, with units managed up about 30% over three years. The pressure on improved productivity will be huge in this business - a 30% increase in volume over three years with a decreasing high cost headcount. The move to drive work offshore will determine the ability of this business to achieve its profitability metrics. Without a move of this magnitude to offshore support, it would be highly unlikely for an offer of this size to achieve a 30% productivity increase. It certainly is a very aggressive delivery plan for this business's data offer, but I believe it is doable.

The voice service line is in a very different state with a much different plan. As mentioned, this service line is seen as one that can aggressively grow. The plan here is to sell a higher value voice service for core telephony and contact centers, targeting companies who need help managing their increasingly old, legacy infrastructures. While this revenue stream has been growing, the plan calls for a major increase in that growth. This growth will come with an accompanying investment in higher end engineering skills in support of selling a higher end offer. You can see the average unit price increases over the planning horizon, not so much related to a price increase per say, but more to do with a higher function offer being sold. You can also see the investment in high cost geography headcount (about 15 per year). This investment will help support the high growth in revenue, but it comes at a cost. The gross margin of 25% is below the businesses overall gross margin and comes despite a large increase in units managed (40%). With a higher value offer, you see this offer will spend relatively less on third party, non-headcount support, which had played a larger role in the more basic service they were selling before. This is a big bet the business is making. The value proposition of the voice offer needs to take hold with clients or the investment will result in lower margins than indicated in the plan, which are already dragging down the overall margin. And the delivery team has a major action to recruit and grow their voice engineering support to make the higher value offer come alive.

The server management offer is planned to be a bit less dynamic than the other two. The basic server monitoring offer is expected to grow at a fairly steady rate, with consistent price points. Headcount growth will come offshore, which will help to drive additional productivity in terms of units managed per FTE. As long as demand stays solid, the degree of difficulty in this offer is much less than the other two.

Table 3 has some fairly detailed assumptions and tracking required, and you will have to assess how you can best accomplish this. It's also very worthwhile to attempt to benchmark your key revenue and cost productivity measures vs. competition. A first way to attempt to do this can be to work with an industry analyst to see if they have this type of data. The best way is often hiring someone from your competitors. These type of metrics (bookings per BDM, units managed per FTE, etc.), are obviously very difficult to get competitive information on, but it's worth the effort to try to understand how you compare to those you are competing with on a day-to-day basis.

It's worth pausing at this point in the book to reflect on the three year plan. There is a lot of information required to complete the type of plan I just reviewed, some of which may be difficult for your business to do. But I do believe evaluating your managed services business' ability to have an effective plan like this is very important. In sum, here are the keys to focus on in putting this plan together:

⇨ "Unit-izing" your revenue and costs greatly helps the planning process and is worth the effort. If this is not easy to do for your business, I would push yourselves very hard to make this happen. There are ways to get creative to get this done.

⇨ Managing and regularly evaluating the productivity of your delivery infrastructure is critical. It's easy to have this sink to the bottom of the list and must be pushed hard by the leaders of the business. Without the leaders pushing this, it won't happen in the way you want it to.

⇨ Don't get too hung up on the accuracy of your pricing assumptions. It can be difficult to predict this, so conservative assumptions are often a good idea. From a revenue standpoint, your bookings and sales personnel productivity are much more important assumptions and great care should be taken to get this to be as accurate as possible.

⇨ The plan I reviewed here assumes a platform investment has already been made, and the productivity it will generate has a track record of happening. Three year planning exercises that require a significant platform investment are even more important than the example I have reviewed here, and of course it's an absolute must to convince yourself and others in the need for the upfront investment.

⇨ The platform should be a prime driver of enhanced capabilities and offer functionality, as well as improved productivity and scalability.

⇨ This three year plan all comes together as you evaluate and plan for the scaling of your business. The plan is all about having the scalability you need from an efficiency standpoint as your revenue grows. Ignoring this equation will not allow your business to take off into flight. Too many others have had terrific managed services business models, have been able to grow well, but haven't been able to deliver efficiently as the growth came. And with that struggle comes a business that will stall. In managed services, a "hope" strategy that says as your business grows you will become more efficient in a manner that will allow you to hit your goals, without planning for it, is a high risk proposition that ends in mediocrity or failure almost every time.

So the business in this example has put in place an aggressive three year plan. The job of operationalizing the plan and managing the business becomes job one at this point. Of course, the three year plan will need to be adjusted as events unfold, but having the overall plan in place should serve as the ongoing navigational roadmap for the business and should be reviewed and updated several times a year. In terms of running the business, monitoring and managing it on a day by day basis, that requires a strong compass and focus. Let's now turn to that, running the business to get it to the point of lift-off.

Chapter 8 - Operational Management

Once you have a three year operating plan in place with revenue and profit objectives and a unit-ized approach to revenue and cost, it comes down to operationally running the business on a day by day basis. In managed services, it's extremely important to focus on several key metrics in making sure you achieve the goals you have set. I know there are many metrics that many different managed businesses use to track their progress, and in this chapter I will outline those that have worked well for me in the past. The metrics I swear by I have bucketed into four groupings: bookings, revenue, profit and client satisfaction.

Bookings

In generating bookings, the first measure to look at is your funnel. As I mentioned in the business development chapter, the metric I like to use is having your funnel of deals with 40% odds to close or higher at three times your bookings objective for the year. The reason for 40% is that many managed services sales folks, like many in sales, get uneasy about "declaring" a deal is 50% odds to close. I've found tracking 40% or higher tends to have the whole process go smoother. It doesn't matter where in the sales cycle a deal is, even if it's early, if it looks reasonably good and you think you have a 40-50% chance of winning, it should be part of this funnel calculation.

When your funnel of deals that are 40% or higher is three times your bookings objective, you are in an optimal state. When you are 2.5-3 times range you are performing OK, it is still possible to "fly the plane" but extra effort must be made on demand generation and qualification to get the ratio up. At less than 2.5 times you are in a red state and it will be difficult to get the plane off the ground.

The second key measure to look at in generating bookings is your close ratio. Again here I only look at deals that are viewed as 40% odds or greater to close, and what your close ratio is of those deals. I have found if you are closing 33% or higher of those deals you are doing very well. If you are in the 20-33% range you are doing OK, it is still possible to get the plane off the ground. And less than 20% is tough, it just causes so much work in the system (business development through delivery), that it is difficult to run the business at that level. The 33% threshold is not a high one, it should be achievable. Of course as your close ratio is lower, you need a higher funnel to be able to hit your bookings objective. I have certainly run successful managed businesses in the 20-33% close ratio range, which is not a huge issue. Below the 20% clip it gets difficult, that is when you are really in a red state.

As an aside, all existing clients have a scope of work and volume associated with their current deal. When you upsell that client, meaning they sign up for more volume and/ or a broader set of services, that is considered to be a new opportunity from a funnel standpoint and a new booking once it closes.

The close ratio you achieve vs. your funnel yields your bookings total, which is one of the four key measures of the business. Ultimately, your bookings total is the key outcome, but the underlying metrics are critical to track. If you are hitting your bookings objective with a weak funnel but a high close ratio, that is likely unsustainable. Likewise, if you hit your bookings objective by having a huge funnel and low close ratio, it puts tremendous strain on the business cost and workload-wise.

Revenue

Your revenue attainment in any given year basically comes down to two key metrics to track closely, that when combined with your bookings attainment (and resulting revenue in year from those bookings, which is typically about 5/12's of the annual contract value of your bookings), give you your revenue for the year.

The first key metric in this category is revenue retention. You start the year with a base run rate of revenue, defined as your previous year's final quarter of revenue time four (or final month times twelve). What you will achieve from those base accounts revenue wise in the following year divided by your run rate yields your revenue retention percentage. As I mentioned, this is a much better metric than pure renewal rate because small accounts leaving obviously hurt less than large ones. Any retention/renewal metric needs to take into account fully the size of the account. Any account that is growing rapidly needs to be normalized in this metric, so a few key growth accounts don't obscure your true revenue retention.

I have found you want your revenue retention to be at least in the low 90's. Below 90% is a red state. The beauty of a managed services annuity business is the "stickiness" of client relationships. When you are in a client and truly managing a piece of their infrastructure, it is hard to get dislodged. The mark I set as the bar for the revenue retention metric is 90%. When you are operating at over 95%, your business is running at a high level. 90-95% is OK, the plane can fly at that level. Below 90% will not work. If you have a smaller business, one key client can put you below 90%. That can happen for a year, but you can't have a high flying managed services business running below 90% revenue retention for too long.

The second key metric for revenue is existing account growth. You enter each year with some clients still in ramp-up mode. It's important to track this during the year. I have found you want to have about 2-5% of your revenue in any given year be from accounts ramping up to their expected volume. Now this percentage can vary widely when you have a smaller managed services business, say less than $10m a year, but it is still a key metric to track. Regardless of the size of your business, less than 2% is a red state. The business is way too stale at that level, over 5% is a very vibrant, growing business.

The revenue you get in-year from bookings, when combined with your base revenue times your revenue retention percentage plus existing account growth, yields your total revenue for the year.

Profit

Calculating your profit obviously starts with your revenue attainment, and then managing your costs closely to achieve the profit goals you have set. Your cost base includes headcount and non-headcount costs to be managed closely, and I like to track two overall metrics.

The first metric, or "key indicator", is around measuring the units managed per FTE (full time equivalent) by technology, as we reviewed in the previous chapter. I have found this to be the best underlying efficiency metric for headcount, and once you set your efficiency metrics for the year, it really is critical to achieve them. I use a percentage of attainment of the units per FTE as the metric. 100% is the goal, and 80% or below is a red state. You will need to average your productivity attainment across the number of technologies you manage. So if you have three key technologies you manage, when you average your attainment across all three, the goal is to be north of 100%. The key here is the planning process we reviewed in the previous chapter, you can't kid yourself with "make believe" metrics. If you've got solid efficiency metrics set for your delivery team, religiously tracking and adjusting as appropriate is critical.

The second key metric is managing your non-headcount costs. You should have a non-headcount cost as a part of your annual cost plan, likely between 25-50% of your overall costs. As you determine your cost plan for the year, a certain percentage of your costs will be non-headcount, and it's that percentage that should be tracked closely. Your percentage attainment of that percentage (non-headcount percent of overall cost) is the metric. And again here, 100% attainment is what you want and 80% or below is a red state. When it comes to non-headcount costs, tracking by vendor is critical. And your non-headcount report by vendor is an important report to track and manage - we will review key reports a bit later.

The third metric in this section is the percent attainment of your SG&A spending goal. This is calculated in a similar way as the others, with 100% being your actual SG&A spend equaling you're the plan you set, and anything less than 80% of your target being a "red" issue. (80% here means you are overspending by 20% over your target). The percentage of revenue you spend on SG&A will vary greatly by the maturity of your business. More mature businesses should see the SG&A percent of revenue be 10% or below. That percentage can get north of 25% for newer businesses that need to invest in sales, tools, etc. in order to grow.

Client Satisfaction

Your bookings, revenue and profit outline the level of activity in the business from sales through delivery, and all that activity results in the client satisfaction you achieve. Of course, it all must come down to having satisfied clients. Your revenue retention depends on client satisfaction obviously, and, in addition, your funnel and close ratio ultimately do as well. References are such a critical part of closing larger managed services deals, as we've discussed. So having clients that are satisfied and willing to be references is key to any managed services business.

For me there are five metrics to watch and manage in this section.

The first operational metric is percent proactive, as we discussed previously. 95% is the target here, if you are remotely managing a client's infrastructure, you should be seeing and opening a ticket via your platform in almost all cases. Below 90% is a red state.

The second metric is percent of service impacting tickets that take greater than eight hours to resolve. It's these issues that take long that can really degrade client satisfaction. You want less than 5% of your tickets to take greater than eight hours to resolve, with <2% being outstanding.

The third operational metric is the percentage of on-time implementation. This is determined by taking the agreed to roll-out plan with the client and what percentage on time you hit. The plan dates can change prior to the beginning of the actual implementation, so it's typically best to use the latest plan dates prior to the first transition. The target here is 90% on time, below 70% is a red state. It's best to adjust the schedule you measure against in extreme situations only, and only when the client agrees to it. This is another metric you can kid yourself on if you want, but it's best to remain disciplined.

The fourth metric is around client specific SLA's, typically for larger clients. These SLA's, which will often come with financial penalties for non-performance, must obviously be closely monitored. 95% of all client specific SLA's is a bar I have used, below 90% is a red state. Above that level you should be OK, with 100% obviously being the goal and expectation in any given time period.

These four above operational metrics I have found to be great indicators of client satisfaction. The fifth is around the view of your overall client satisfaction, the collective "temperature" of your client. Of course, there are so many things that can impact how happy any particular client is. You must have a process and a vehicle for gauging client satisfaction, and then putting resulting plans in place to remedy any key dis-satisfiers. We reviewed in the Service Delivery section my view of client sat surveys. I am not a fan for managed services of elaborate client satisfaction surveys, in fact, I am a particular non-fan. I like to get from the client decision maker a grade of A-F, or a numeric scale of 1-5. And as I discussed previously, the goal is B or B+. Anything in the C range or below is a red state from an overall standpoint. So, fully tracking client satisfaction and achieving a B or better overall rating is the goal. This overall rating is the fifth metric in this section which should be tracked. We reviewed in the client management section just how critical the day-to-day interfaces are in staying on top of and reacting to key client issues and opportunities. All that effort needs to be reflected in the grade each client gives you and your overall cumulative grade.

So in this chapter we have reviewed the key metrics to track in bookings, revenue, profit and client satisfaction. As I've said, I am a big believer in having a handful of metrics to track the "underlying" health of your business, and we have reviewed in this chapter those I think you should seriously consider being a part of your management system.

Chapter 9 - Reporting

As you run your managed services business on a day by day basis, there are a series of reports you will need to monitor closely. This chapter outlines the key reports I have used which I've found very helpful.

Business Development:

Funnel Report - Nothing out of the ordinary here, just your basic funnel or pipeline report. I have typically produced these weekly, no less than bi-weekly.

Client Geo. Target Close Odds ACV TCV Contract Length Comments/Next Steps

I break out the deals that are over and under 40% odds to close. The close rate of any deal that's greater than 40% odds to close (at any point in sales cycle) is calculated from this report.

This report also will show which accounts have closed, what your close ratio is, and give you your bookings total.

So a report like this is critical to tracking and managing the health of the funnel, how effective you are in closing deals and the ultimate measure of your total bookings.

Revenue:

There is no substitute to me for monthly revenue report of every single client. It will give you a clear sense of the health of your client base. Any potential instability must be dealt with with great urgency. I view it has a must to stay on top of this report.

Revenue																	Year 1																Year 2		
Client	Geo.	Ctrt End	J	F	M	A	M	J	J	A	S	O	N	D	Yr	J	F	M	A	M	J	J	A	S	O	N	D	Yr.

I've shown above a monthly view that would get updated with actuals each month. I typically break out the client in any given year into base clients (those you entered the year with) and new clients sold within that year.

This report enables you to calculate the all-important revenue retention percentage discussed earlier. It also is the report used to calculate your account growth in-year totals and percentage, as well as your base revenue going into any year. With accurate contract end dates, you can also calculate your total revenue under contract.

One other metric that is very beneficial to track that you can get from this report is your ratio of total revenue under contract to your annual run rate of revenue. As I mentioned, this by client revenue report outlines the contract end date of each client, along with the monthly billing, which enables you to calculate the total revenue you have under contract quite easily. That number divided by your annual revenue run rate at any point in time gives you that ratio. A 2:1 ratio means your average client has two years left on their contract, which is an outstanding ratio. This is not a metric you need to be all over on a day by day basis, but it's very good to calculate and track this at least quarterly to give you a feel for where your backlog is and how predictable your future growth will be. We'll talk a bit more about this in the Valuing a Managed Services Business chapter.

This report not only will generate many key metrics, it also gives you a great "eyeball" test on the health of your business. I make it a point each month to look at every client on the list and assess if things are going as planned, what key contract renewals are out there or coming up, etc.

Client Management:

Gauging the temperature of your clients each month is certainly important as well. I typically will do this for all clients that make up 70-80% of the total managed services revenue stream, I don't track smaller clients in reporting like this. What I will do in addition to the report outlined below, is have a monthly "Top 10" report of the ten biggest clients and then have a more basic report for the rest of the clients.

For the Top Ten report, it's typically a three to five page Power Point presentation. Key sections in this report are an executive summary (5-10 bullet points, including an assessment of the clients overall satisfaction with an A-F grade), a scorecard chart on revenue, SLA performance, upsell opportunities and new deployments, and a current issues risk review with mitigation plans.

For all clients in the 70-80% of revenue, outside of the "top 10", a report such as the following provides for me the right information to insure all awase.

Client Ann. Rev. Ctrt End Client Grade + Temp. Retention/Upsell Activity Issues Actions

I typically like a few short sentences to describe the last four columns in the above report. The number of clients on this report should be equal to the ones you expect your team to be interacting with on a very regular basis. Of course there is no need to do this report for clients that need to be "low touch". As mentioned also, I typically like to have 70-80% of total revenue represented from the clients on this report.

The satisfaction of your client base can be easy to spend less time on as you look to grow revenue and get more efficient and maximize profit, but it must be a focus. It's critical to have references and to keep your revenue retention above 90% at a rock bottom, 95% being the target. It's simply an area that must be invested in. I feel a report such as the above can keep your entire organization on its toes and focused on current client satisfaction, as well as be used to tightly measure the effectiveness of the people you have directly responsible for ongoing client management.

Headcount Costs:

Certainly closely managing costs are an integral part of successfully running a managed services business. I have found it very helpful to actively track both your headcount and non-headcount costs. The headcount cost report format I have used is a simple one as outlined below, broken into cost and expense categories.

	Headcount	Total Cost
	J F M A M J J A S O N D	J F M A M J J A S O N D

Cost
Dept 1 Total
Individual   1
             2
             3
Dept 2
Individual   1
             2
             Etc
Etc

Expense
Dept 1
Individual   1
             Etc.

When you use outside contractors in place of headcount, I prefer to track that in the headcount report as opposed to a non-headcount report.

As you are looking at your cost metrics (ie., units managed per FTE), it's important to have a report like this to analyze where you are and what the headcount plans are in each area. Even though listing each person can be a burden, it's good to have that level of granularity. Due to the sensitive nature of the data, only a few people in the organization can have access to the detailed information.

Non-Headcount Costs:

Managed Services business often vary greatly in how much non-headcount costs, typically mostly third party, are involved in their client solutions. It's almost always tangible enough to warrant very tight management. Certainly tracking the actuals vs. planned spend is crucial. Here is the basic format of a report that can be used to track these costs monthly. It should list out each vendor used within categories (ie., third party maintenance, contractors, etc)

	Planned Spend													Actual Spend													
Cat.-Vendor	J	F	M	A	M	J	J	A	S	O	N	D	Yr	J	F	M	A	M	J	J	A	S	O	N	D	Yr	Comment

It's important to include in the report a comments section on the view of the spend and any action that is required. Is it in line with the plan, what is being done to reduce the spend over time, is there an issue with what's being charged, are all items that should come out in the comments. Not all non-headcount spend may be third party costs, there will also be internal non-headcount costs that should be tracked. There should be clear ownership in the organization on non-headcount cost management.

Delivery:

Each managed services business needs to decide which delivery metrics are most critical and track accordingly. As I've mentioned, it is best to clearly identify a few critical metrics and managing tightly. For whatever metrics you choose to measure, a summary report like this could be used:

	Performance											
Metric	J	F	M	A	M	J	J	A	S	O	N	D

% Proactive

TTR> 8 Hours

MTTR

Etc. (other metrics in client SLA's or other key delivery metrics)

Any client with custom service level agreements should be reported out monthly as well.

Implementation:

Most managed businesses have a very active onboarding effort with multiple clients. It's a key activity to shorten time to revenue, and successful onboarding, along with the first six months of day 2 operations, typically determine the client's satisfaction though the life of the contract. It is of utmost importance to have a well-documented process and supporting metrics. A basic implementation report like this can be used to track:

Client	PM Lead	% Complete	% On Time	Comments

For larger clients, the full roll-out schedule by site, by date, with planned/actual cutover really should be reviewed. Percent on time is the key metric here, and it can be a difficult one to manage. The dates set out at the beginning of the project, agreed to with the client, are the dates to hit to figure out the on time percentage. Of course, clients can be the cause of date shifts. Careful inspection must be done to insure any plan date changes are truly because of the client.

The reports reviewed in this chapter are fairly basic and are consistent with the key metrics identified in the previous chapter. Once you have your three year strategic operating plan, identify the key metrics to track, and insure you have clear reporting, you are well prepared to run a very tight managed services ship.

It's a fine line between burdening your business with too many reports and running a business in too loose a fashion. The number of reports outlined in this chapter represents the balance I have found works for me.

Chapter 10 - The Managed Services Inspection

As you can probably tell, I feel it's extremely important in running a managed services business to have a detailed plan and track it, and then adjust in a dynamic fashion. The fact that this is a high fixed cost, low variable cost business requires strong planning and tight management. The previous few chapters talked about strategic planning and the operational management of the business, including key metrics to track and the types of reports you need to effectively run the business. But holistically, how do you know how you are truly doing? Even with a condensed set of metrics, how can you tell when you examine your performance in total if you are in satisfactory, exemplary, or in poor shape overall?

This next chapter attempts to answer that question with what I call the Managed Services Inspection. I think of it similar to a detailed inspection done of a jet in the hanger once a month (vs. daily between flights). This inspection will help to clarify when a business is not ready to take flight. The business that fails this inspection will very likely not be successful and needs to readjust and "re-plan" before it attempts to take flight again. This inspection also reveals the minimal score you need to be successful, to be able to take flight. And it will also identify when your business is in true optimal condition, ready to take flight in a highly effective and efficient manner.

This inspection is designed to enable you to look beneath the high level results of revenue, profit and bookings to understand how truly effective the business is and how well positioned it is for success going forward based on the performance of the key underlying metrics you are tracking.

This inspection grades how well you are doing to achieve your financial plan. It makes sense that before we get into the inspection, we take a quick look at what targets for the financials of your business are sufficient. It is very difficult to be too prescriptive about this, as there are so many factors that need to go into your financial plan. But I thought it would be useful to outline what types of growth and profit you should be looking at given the size of your business. If your three year business plan does not get you to the levels of revenue and profit I outline below, you should definitely take a step back and insure you have the right plan. There certainly may be good reasons for it, but you just have to realize that your competitors most likely have a plan to hit numbers in the ranges I have below.

Of course, there are many factors that go into setting your goals, including the size of your business, the growth rate of the market you're in, the investment that's required, your competitive positioning, as well as many others. I have set out below a high-level view of revenue growth and profit objectives I feel should be obtained.

One of the most critical decisions for any managed services business is to declare the rate of revenue growth you're looking to achieve and having your plans support that rate of growth. Set the growth too high, too unrealistic, and you run the risk of spending too much, ramping up too quickly, and deflating the business. If you set it too low, you run the risk of being competitively de-positioned and not having the scale and momentum to compete effectively. I have outlined below the range of growth that is optimal based on the size of your business, obviously the smaller the business the greater the rate of growth should be, and growth in excess of that is a very good thing!

Size of Business	Growth Rate
< $10m	25-50%
$10-50m	20-40%
$50-200m	15-30%
$200-500m	10-25%
>$500m	5-20%

In terms of the profit of your business and what sort of margin should be targeted, I have found that is less dependent on the size of your business than revenue growth. It tends to be more dependent on where you are in the maturity of your business. I don't think there are any strict rules on what margins you should be expecting, in some ways it's a decision each business needs to make on its own. However, the profit you target does impact the price you will charge, which in turn will impact your market competitiveness.

As you determine your profit goals, what costs you put into "cost of goods sold" (COGS) vs. expenses in "sales, general and administrative" (SG&A), needs to be considered. There is definitely some subjectivity into what costs you say are part of your direct costs of providing the service and what costs are not and go into expense. If you want to inflate your gross margin, you can lean towards moving more costs into SG&A. However, there is no way to inflate your EBITDA, earnings before income tax and depreciation. In many ways therefore, EBITDA is the best true measure of your profitability.

In general, managed services businesses average gross margins in the low 30's. Most outsourcing businesses are in the mid to high 20's in gross margin and cloud businesses are often in the 40's and 50's, with higher SG&A expenses. I have certainly seen many cloud businesses with inflated gross margins, inflated SGA expense and very little, if any, true bottom line profit. In fact, according to the latest TSIA Cloud 20 study, the average gross margin for a cloud provider is 57%, yet their bottom line profit is -1%. Your bottom line profit, EBITDA, is the true profit your business is generating. If you are at 25% EBITDA or better, you are doing extremely well. With the investments you continually need to make and the competitive nature of this type of business, to achieve at 25% EBITDA is an outstanding accomplishment. Most managed businesses are not at that level, most tend to run in the 10-15% range. A managed business operating at below 10% EBITDA really needs a strong plan to get into double digits. There is some owner of that business somewhere that would rather put their money somewhere else at some point if you can't at least make a 10% bottom line profit.

So let's now turn to the inspection and how it works and calculates its conclusion. As I said, the Managed Services Inspection covered here is designed to help you truly understand and grade how you are doing in driving top and bottom line results.

You've built a jet plane in your managed services business and have set off to hit your strategic objectives, and once a month you should be sending this jet plane into the hanger for a deeper inspection than you give it on a day by day basis. This once a month inspection and grade will outline which of the following three states your business is in:

1) You are ready to soar towards your objectives in a truly *optimal* fashion.

2) The jet is ***ready to fly*** and fly well, there are some items needed to improve and focus on or the next inspection may not go so well.

3) The jet needs to be ***grounded***, there is a need to remedy a key part or parts of the business before you can soar towards your stated objectives.

There are four quadrants to the inspection: Bookings, Revenue, Profit and Client Satisfaction. The inspection measures key metrics below your high level results and determines a numeric grade for each quadrant and in total for your business. Each quadrant has two to five key variables to inspect. All four quadrants are of equal importance, each with a total potential score of 300. Each variable is graded into one of the three categories outlined above: optimal, ready to fly and grounded. Each variable has a priority attached to it from one to three, and your score is determined by the grade and the priority. The table below shows the points for each variable:

Grade	Priority 1	Priority 2	Priority 3
Optimal	200	100	50
Ready to Fly	100	80	40
Grounded	0	0	0

So let's look at each quadrant and how the scoring will work for each.

Quadrant 1 - Bookings:

There are two variables in the bookings quadrant: funnel size (of deals >40% odds) and close ratio of those deals.

Funnel Size (relative to size of bookings target) - Priority 1

3x	Optimal	200 points
2.5-3x	Ready to Fly	160 points
<2.5x	Grounded	0 points

Close Ratio - Priority 2

>33%	Optimal	100 points
20-33%	Ready to Fly	80 points
<20%	Grounded	0 points

I truly believe the funnel size is one of the most important measures of how fundamentally strong your business is. The process must have strong integrity to it, and with that you will have an excellent gauge on your future success.

Quadrant 2 - Revenue:

There are also two variables in the revenue quadrant, revenue retention and account growth.

Revenue Retention - Priority 1

95%+	Optimal	200 points
90-95%	Ready to Fly	160 points
<90%	Grounded	0 points

Account Growth - Priority 2

>5%	Optimal	100 points
2-5%	Ready to Fly	80 points
<2%	Grounded	0 points

Like funnel size, your underlying revenue retention is just such an important metric to achieve. Instability there is almost impossible to make up elsewhere, therefore it's a "priority 1" metric. Smaller businesses can lose one big client and fail that metric easily, so if you are sure that one client you lose is the last of your at risk large clients, you could manually override a "grounded" grade in revenue retention if you are highly confident there are no other large clients at risk.

The first two quadrants are basically inspecting and evaluating the underlying factors behind your revenue performance - as your base run rate entering the year, factoring in revenue retention and account growth in the following year, plus the revenue you get from new bookings, which is based on the close ratio and size of your funnel, gives you the revenue you will recognize in the year.

Quadrant 3 - Profit:

There are three variables to look at in profit in this inspection. The targets for each of the three - units managed/fte, non-headcount percent of total cost and SG&A percent of revenue - are based on the profit goal set for the business and the targets for each variable we discussed in the previous chapter.

Units Managed Per FTE - Priority 1

100%+	Optimal	200 points
80-100%	Ready to Fly	160 points
<80%	Grounded	0 points

Non-Headcount Percent of Cost - Priority 3

100%+	Optimal	50 points
80-100%	Ready to Fly	40 points
<80%	Grounded	0 points

SGA Percent of Revenue - Priority 3

100%+	Optimal	50 points
80-100%	Ready to Fly	40 points
<80%	Grounded	0 points

All three variables in this quadrant are important, but I place significant weight on the underlying operational efficiency of your delivery team. And as we've discussed, I do believe units managed per fte, calculated with integrity, gives you the best measure of now efficient you are and how ready you are to scale.

Quadrant 4 - Client Satisfaction

There are five variables in this quadrant, the overall client satisfaction grade your clients give you being most important, a priority 2, and four supporting operational metrics, each a priority 3.

Client Satisfaction Grade - Priority 2

A–B+	Optimal	100 points
B–B–	Ready to Fly	80 points
C+ or lower	Grounded	0 points

Percent Proactive - Priority 3

95%+	Optimal	50 points
90-95%	Ready to Fly	40 points
<90%	Grounded	0 points

Percent Tickets > 8 Hours to Resolve - Priority 3

<2%	Optimal	50 points
2-5%	Ready to Fly	40 points
>5%	Grounded	0 points

Percent On Time Implementation - Priority 3

90%+	Optimal	50 points
70-90%	Ready to Fly	40 points
<70%	Grounded	0 points

SLA's Large Clients - Priority 3

100%	Optimal	50 points
90-95%	Ready to Fly	40 points
<90%	Grounded	0 points

The maximum points from the four quadrants and twelve variables is 1200 points. Overall if the score is 1140 or higher, your managed business is in a strong state to achieve the goals you have set. A score from 1000 - 1140 is a strong business that is ready to fly and on the doorstep of being in an optimal state. And a score of less than 1000 I would say is in a red condition needing improvement on the identified "grounded" variables before being ready to fly.

This inspection exercise is designed to evaluate how your business is doing vs. the goals that have been set, it's really not designed to evaluate how you are doing vs. your competition. Your overall goals must be set in a way that will give you an edge over your competitors. I believe this inspection is an excellent way to understand the sustainability of the business results you are seeing. Even with one variable getting a "grounded", it should be cause for concern and immediate focus. Ignoring any of these key variables will most likely come back to bite any managed business.

To help illustrate the inspection, I will show two "real life" examples.

The first example is a point in time in one of the managed businesses I ran, this one at Aimnet. The inspection of the business at this point in time shows an overall growing business but with some areas that clearly raise some questions or require action.

Bookings Quadrant:

Funnel Size -	160 points	2.8x+
Close Ratio -	80 points	23%
Total -		240 points out of 300 points

At this point, the business was growing fast, though it was overly dependent one partner. The funnel was just south of an optimal state, and the close ratio was pretty good, good enough to hit the bookings target. There was more churn in the deals than optimal, creating more pre-sales work than we'd ideally like, but overall the demand generation and closing was strong enough to hit the bookings objective

Revenue Quadrant:

Revenue Retention -	160 points	92%
Account Growth -	80 points	7%
Total -	240 points out of 300	

There was a lot going on in the account base at this time. There were several larger, long-term clients that had left, which had revenue retention at a concerning level, though still adequate. In addition, there were several existing accounts that had grown quite substantially recently which enabled the existing account growth to be strong. While this quadrant was fairly strong, 92% revenue retention is certainly one that bears watching. There's not a lot of room to have retention lower than the level here to be able to hit your goals.

Profit:

Units Managed Per FTE -	200 points	100%+
Non Headcount % of Cost -	40 points	96%
SG&A % of Revenue -	50 points	100%+
Total -	290 points out of 300	

The profit of the business was good at this point. The underlying efficiency was especially strong at this point, the business was very ready to scale. A few clients required us to partner for some break fix services which increased our non-headcount spend, but it wasn't a big deal. The business was ready to grow efficiently.

Client Satisfaction:

Client Satisfaction Grade -	80 points	B-
Percent Proactive -	50 points	97%
Percent Tickets > 8 hrs to resolve -	50 points	2%
Percent On Time Implementation -	40 points	77%
SLA's Large Clients -	40 points	95%
Total -	260 points out of 300	

The client satisfaction was pretty good, but there were a couple of concerning areas. There were two accounts that were problematic at this point in time, which lowered the overall client satisfaction grade and the performance on service level agreements. The other issue the business was having was delays in rolling out new client implementations. We were struggling at this point on how much of the delays were self-induced versus client related. Nevertheless, it was an issue.

Grand Total - 1030 points out of 1200 - a "ready to fly" grade

The overall grade is a good one, and the business was generally strong. But there were key focus areas that very much needed attention. The top line revenue growth was a little less than we were hoping for. When you looked at it, the issue was not really demand generation or the ability to close, it was more related to the existing client base in that a few accounts that left had impacted the base revenue making it that much more difficult to grow top line. We were achieving our profit objectives due to strong gross margin performance and the underlying scalability of the delivery model. We had a few accounts that had left, we also had a few larger accounts that were current customers where we were having issues. It was impacting our overall client satisfaction and SLA performance, as noted above.

The business at this point had some real strength in the demand generation and bookings area as well as the scalability of the business. But the existing customer base was not being managed as well as it should be. To get this business to fly in an optimal state, an increased focus on our existing customers, especially the large ones, was required. The zeal for the business to close new deals and to ensure we could scale effectively had overshadowed a bit ensuring our core customers were being taken care of completely. We were 30 points from the "mendoza line" of 1000 and 70 from an optimal state, and we really had to focus on the current client base, while not taking our eye off the ball on the other areas of the business, to get to an optimal state.

The second example is a point in time at AT&T. As AT&T was just starting to really focus on managed services, the inspection of business at this point shows a booming business, but one that did have some issues that were holding it back.

Bookings Quadrant:

Funnel Size -	200 points	3X+
Close Ratio -	80 points	25%
Total -	280 points out of 300 points	

The core value proposition was very strong at this point. A bundled managed router offer was taking off with both enterprise and mid-market clients. The business was growing fast and the close ratio was fairly strong (though later the close ratio would lower and cause some issues.)

Revenue Quadrant:

Revenue Retention -	200 points	96%
Account Growth -	80 points	3%
Total -	280 points out of 300	

The account base quantity wise was growing rapidly, there was not a lot of attrition at this point. There were a lot of new customers early in their three year deal cycle. In addition, the newness of the client base did lead to only moderate account growth. The low account growth was really not a concern at this point, we figured it would grow over time, which did prove to be true. As time went on the revenue retention did lower from this level.

Profit:

Units Managed Per FTE -	160 points	94%
Non Headcount % of Cost -	40 points	91%
SG&A % of Revenue -	40 points	82%
Total -	240 points out of 300	

The growth of the business hit a bit sooner than the business was ready for from a delivery standpoint. We were not hitting our gross margin or EBITDA targets at this point in time. In the units managed per fte area, there was one technology area that really dragged down the metric, with too many dedicated people and weak platform support. Our non-headcount spend had grown more than we expected due to sub-sourcing our low end support to a small managed services company, which wasn't very concerning to us as the margin on those deals did exceed what we were doing internally. The SG&A was a weakness on two fronts. The sales cost was high, which we somewhat proactively did to maximize our growth. But what really impacted this area was the ongoing investment in our platform. It was costly, too costly, and had a very negative impact on our profitability. We were getting to be pretty close to a "grounded" zone on SG&A.

Client Satisfaction:

Client Satisfaction Grade -	80 points	B
Percent Proactive -	40 points	93%
Percent Tickets > 8 hrs to resolve -	40 points	4%
Percent On Time Implementation -	50 points	91%
SLA's Large Clients -	50 points	100%
Total -	260 points out of 300	

The client satisfaction was pretty good. While this business had some growing pains client sat-wise prior to this, the core client satisfaction was good at this time. The on time implementation was actually especially good given the tremendous ramp up that was occurring, which helped revenue recognition in addition to client sat. We as a business were learning that those first three to six months were incredibly important for the client satisfaction and relationship over the life of the contract.

Grand Total - 1060 points out of 1200 - a "ready to fly" grade

This was a strong business in most areas, with a strong foundation of growth and good client satisfaction. But the cost and profit of the business held it back from achieving an optimal state, with SG&A missing its target by an alarming margin. The focus needed to be on the scalability of the delivery infrastructure. The actions were primarily on two fronts - fix the productivity issue we had in one technology area and get our platform costs under control. This business did get to an optimal state, but not without a fair amount of pain. We needed to scale back platform requirements and get the platform team under control from a headcount standpoint. As we fixed the scalability issue in one of the technology areas, the business was in much better shape to drive the needed profit as growth continued.

Chapter 11 - Valuing a Managed Services Business

It's one thing to set the right growth and profit goals for the company and manage it closely, inspecting key metrics to insure you are on track. It's quite another when you get involved in any merger and acquisition discussions (M&A), either looking to acquire a managed services company or looking to sell your managed/cloud business. I have been involved in many of these discussions and evaluations in my career, and I have come out of those experiences with some definite thoughts on how best to approach this.

The valuation of a managed services business in terms of revenue or EBITDA multiples is obviously highly dependent on market conditions at the time. I'm not going of focus on what multiple you should be paying or selling for, but more on how can you best predict the future revenue and profit of a managed services business to base a valuation on. When buying a managed services company, you obviously need to project out what the profit and loss statement will look like over the next three to five years. How to best do that if you are buying a managed services company, and how you should evaluate yourself if you selling, is what I will focus on.

When you're buying managed or cloud business, one of the things you will look at and evaluate is that future P&L statement. It will tell you what the selling business thinks their business will look like, hopefully based on a set of facts that can be supported. From a valuation standpoint, I look at the projected P&L of a managed services business, understand the dynamics and the rationale, and then pretty much throw it out. It inevitably is not so much the likely future financial plan, but rather the maximum revenue and profit this business thinks they can achieve if everything goes right, plus a little bit of luck.

What I like to do is really get under the numbers of the business the previous two to three years and generate a future P&L statement based on the metrics the business is achieving. So, basically, in doing diligence on a managed business, I want to recreate tables 1-3 from pages 113-116 looking backwards at the previous three years. Of course, many businesses will not have reported on their business that way, but it is beneficial to recreate the past three years in that format. Most of the data the business will have. You certainly need headcount data by function and by technology, as well as unit pricing information. Some businesses may not have good unit pricing information, but you can reengineer the financials of the business via what they have sold and are running to come up with representative unit prices. For me, there is no way I want to make an investment in a managed business without having those three tables filled out for the past three years, even with some figures being only estimates.

The two things I try to focus on is the revenue projection based on recent performance and how the profit and efficiency of the business changed as the revenue changed. It's just so critical to evaluate whether the demand generation of the company is working in terms of new sales, clients and revenue, as well as determining if this business is able to scale - using the results of the recent past to derive the future projections.

From a revenue standpoint, I want to stare long and hard at the revenue report we reviewed earlier and shown below.

Revenue				Year 1													Year 2												
Client	Geo.	Ctrt End	J	F	M	A	M	J	J	A	S	O	N	D	Yr	J	F	M	A	M	J	J	A	S	O	N	D	Yr.

I want this report back for the previous three years and that gives me just about all I need to know from a revenue standpoint. Here you see how new sales truly manifest themselves into revenue, what the revenue retention is truly like, and to what extent existing clients are growing. It is very helpful to evaluate if any of these factors are highly dependent on a few key clients, or is what's happening in the business more spread out among many clients.

One key statistic I want on a running basis for each month is the total revenue under contract at that point in time, which we discussed briefly earlier. With each client's monthly revenue listed along with a contract end date, you can calculate the remaining revenue you will see from that client. At any point in time you should show the revenue remaining under contract, and it is helpful to keep a running total on what the ratio is of revenue under contract vs. your current annual revenue run rate. That is another way of showing how much time left the average client has in their contract. That ratio is a very good indicator of the revenue predictability of the company. Obviously the closer to '1' the ratio is, the worse the revenue stability of the business. We've discussed that the average client contract is three years. The more your business is oriented towards outsourcing, the more five year deals you will get, and the more it's oriented towards public cloud and SMB the closer to two years (or even one year) it will be. Many public cloud businesses have one year deals as a standard. I want to acquire, and be selling, a managed business that has three years plus as its typical contract length. And I want that ratio of total revenue under contract to annual run rate revenue to be 1.5:1 or greater. At a 1.5:1 ratio, that basically says your average client has one and a half years remaining on their contract. If you are closing three year deals, 1.5:1 would be roughly what I'd consider an average ratio. The more success you have had in closing deals in the recent past, the higher the ratio will be. A ratio of 2:1 for a business that is doing three year deals is very good. That means you have had a steady influx of new deals. The highest I ever had that ratio was 2.8:1 at Avaya, where we sold a number of five year deals and the total under contract was very strong relative to our revenue run rate.

One critical area this report will also show is how dependent the future projections are to the biggest clients. If growth was highly dependent on a couple of key large clients, extreme inspection of the top deals in the funnel is required to insure your estimate of future revenue is correct. If revenue retention has been strong, but over the next year there are very large clients up for renewal, that must be inspected closely as well. That is especially true if there is not a long track record of how large deal renewals have gone. The beauty of this report is there is just no way to hide anything. The retention you project must be based on the past, and the existing account growth is the same way. And if there are certain existing clients that will swing these results, those must be inspected and in most cases that means talking live to the decision maker at those few key accounts.

In projecting out the next three years retention, account growth, and new revenue, I believe it should be done in the revenue report format above. Each account should have a forecasted revenue stream, making overall assumptions on retention and existing account growth that need to manifest themselves in the by account forecasts. The forecasting for the largest five percent of accounts will need to be done with a high degree of diligence. And then you need to forecast the new business in the report, showing the size of the client, contract length, etc. And if the business has been dependent on a few very large new client wins, the funnel needs to be inspected to assure you are comfortable with any new business projections.

As you complete filling out that report based on past results, a strong picture of the future revenue of the business should emerge. Any uplift on that revenue projection you come up with should only be done with extreme diligence. Obviously what you'd like to see is not only a growing revenue stream, but also a growing revenue under contract metric.

Selling a managed business under this revenue microscope is not fun, it shows any warts pretty clearly. But no managed business is without issues, you just need to be able to insure your issues are removable and/or "overcome-able". So when is the best time to sell a managed business? To me it's a pretty simple answer, when you can go through this revenue inspection and projection in as strong a way as possible. Given when key clients have come on board and when others are leaving, or have just renewed, are all key factors to consider in going through this level of diligence.

The other key part of doing diligence in buying a managed business is projecting future profit. I really like to evaluate the future profit based on the efficiency shown in the business over the past two to three years. Any business you are acquiring or selling has a certain revenue trajectory it has seen in the recent past, and I want to know what the underlying efficiency of the business was with that trajectory. Hopefully you are buying or selling a growing managed business, and in that light, it is critical to be able to prove the business can scale from a delivery standpoint.

There are "soft" things that must be evaluated in terms of how scalable a managed business is delivery-wise. You must inspect the platform and processes - does the platform drive and reflect the core processes of the business, does it effectively filter out unneeded tickets, does it do an effective job correlating data and making your operations team as efficient as possible are all key questions that must be evaluated. The process documentation is certainly key as well. A scalable managed business is not one dependent on a handful of tier 3 / 4 engineers to come in and save the day time and time again. That is basically the definition of non-scalable. When you have a strong foundation of ITIL-based processes defined and a platform that drives that, you should have an efficient and scalable delivery capability. And that should manifest itself in your results.

In evaluating the scalability of the business, the hard numbers are just as important as the soft items just mentioned. What happened to the core efficiency metrics of the business while the revenue was growing is what needs to be determined. The units managed per fte by technology - how did that perform during revenue growth? The non-headcount costs and SG&A percentage - what happened to those metrics as the business was growing? The future efficiency and subsequent profit you project given the growth you are forecasting is a critical assumption, and the more it's based on facts the better. As your revenue increased by a certain percentage, your underlying efficiency grew by a certain percentage as well. Your future projections of efficiency and cost should be based on what happened in the past. A "hope" strategy of as the revenue grows we can scale, an "all we need is top line growth to scale" view is very dangerous. There is no substitute for proof. The ability to deliver high quality support in an efficient manner as you grow rapidly can only be known by doing it. Most managed business DO NOT continue to deliver quality support in an efficient manner with rapid growth. That is a fact. Typically one or the other or both suffer. It only takes one or two clients where you have delivery issues and/or whose expectations are not being met to tank your ability to scale.

It's a high bar to have a managed business with strong growth projections, based on the recent past, along with an ability to scale efficiently with strong delivery quality, all of which being highly "prove-able". If I had to pick one thing, it's certainly having a proven ability to grow. There's not much you can do without that. A flat revenue managed business that is being acquired because you are confident your go to market and distribution leveraged correctly will result in growth is a big bet. I would not pay a high multiple in that scenario, that's for sure. If you have a managed business that is growing and will grow, but has not proven its ability to scale yet, is a much better situation than the reverse. In that case, the confidence you have in improvements being made in the platform, processes and or delivery leadership will be paramount in your decision making on whether to pursue the acquisition or not.

Chapter 12 - Looking Back

It's interesting to look back at the history of managed services and how it evolved into the market it is today. There is no one certain event that spawned the managed services industry. Back in the sixties you had EDS and their facility management and time sharing business. IBM also entered the computing time sharing business, but they were out of it by the early seventies. I consider the real birth of managed services to be in the eighties.

1980's

IBM and GE were the big name heavyweights in this industry in the eighties, but Telenet and Tymnet were major players as well. Telenet and Tymnet were packet-switch network providers who provided managed network services. They started in the seventies, and grew rapidly in the eighties, providing dial up remote access to computers for corporate users. GTE and then Sprint owned Telenet in that decade, while McDonnell Douglass owned Tyment for much of the eighties before selling it to BT at the end of the decade.

GE Information Services (GEIS) started in the sixties as a time sharing company, and by the eighties it had migrated to become a value added network service and EDI (electronic data interchange) company. This managed network and inter-company electronic communication business grew rapidly, especially its EDI business, later in the decade.

IBM started its IBM Information Network in 1982, which is when I started with the company. It provided remote computing services, value added network services, and by the middle of the decade, EDI and Email services. Looking back at the remote computing business, it was not much different than the cloud offerings of today (just a lot less successful from a top line standpoint).

It's hard to imagine now those pre-Internet days when clients were looking for network solutions that carriers, namely AT&T, couldn't deliver. IBM and the others rapidly built out private networks for clients to leverage to run their own internal networks. Unlike other vendors in the space, IBM built out its network based on its own proprietary SNA technology.

To me, the first managed services 'boom' was remote dial up access and that was followed by the rapid expansion of electronic messaging in either standard data (EDI) or unformatted (Email) fashion. In the late eighties, we at IBM saw ourselves in a fight with GE, Sprint/Telenet and MD/Tymnet. The industry was growing at about 15% at the time according to industry analysts. We were a $500m business growing at about 30% making about a 20% gross margin, making most of our money on slower growing network services and much less in our faster growing messaging businesses. Our goal was to dominate intra and inter-company communications. At the time, we actually felt most threatened by Sprint, who had acquired Telenet. While GE was very strong in EDI, their core network business was weak and a major competitive disadvantage. Sprint Telenet had a larger market share than we did and having the backing of a carrier was something that concerned us. However, we felt our

SNA network was a major competitive advantage, and the IBM client base who depended on their mainframes would provide a great foundation to grow from. A key strategy we had, as I briefly reviewed earlier, was to give away, more or less, the IBM mainframe connectivity to our network and therefore making it easy to add remote client sites either by dial up or dedicated lines. While our competitors had strong X.25 networks that were very flexible, the performance and security of SNA, and the presence of IBM mainframes in corporate enterprises, gave us what we felt was a big leg up. And for a while it did.

1990's

As the decade changed, the landscape of managed services changed dramatically as well. I have a few industry analyst reports from back then where I saw the term "managed services" used for the first time. Client server began to expand in the IBM dominated enterprise market. The network routing required for these infrastructures was a definite threat to the IBM-dominated SNA network business. I can recall many a meeting at IBM where the future of SNA and this relatively new company called Cisco was discussed. We felt at the time our ATM network technology would win the day, overtures from Cisco were rebuffed and our investment in next generation IBM network controllers was increased. The IP revolution was right in front of us, we just couldn't see it. Looking back, it's easy to see how off base this truly was.

Early in the decade, two major trends were taking hold. IBM's Global Services business, known as ISSC at the time, which had begun making major inroads in data center outsourcing in the eighties, saw a need for network outsourcing. IBM and EDS began to battle in network as well as data center outsourcing. By the middle of the decade, major carriers such as AT&T, MCI and BT had entered the network outsourcing market.

Another even more significant trend started to occur, the emergence of Frame Relay as a flexible, efficient alternative to dedicated "leased lines" from the carriers. AT&T, BT, MCI and Sprint among others rolled out the first frame relay services and began to offer various forms of managed services around their new network service. IBM began to offer a Frame service as well. But it was way too little too late, the tide was turning. The world of managed services was tilting the carrier's way. Client server and associated frame relay support changed everything. GE's strength in EDI was not enough to offset its network weakness and it struggled (and reached out to IBM about a merger at one point). Sprint Telenet and Tymnet, now a BT company, also struggled to find the right niche with their new carrier's portfolio.

The IBM Information Network became Advantis in 1992 (merging with Sears network service business) and was renamed the IBM Global Network shortly after that. The major growth began to come from larger network outsourcing deals. By 1995, the IBM Global Network was a $1.5b business at a 25% gross margin. While the core network service was increasingly becoming a competitive disadvantage, the value added of managing a client network was becoming a major advantage. However the core network was a boat anchor for the business. The investment required was significant. The costs the carriers offered clients with their frame relay solutions were less than we could offer. Mainframes and SNA were on the way out and IBM decided by 1995 it was time to punt, ultimately selling the business to AT&T in 1998.

As IBM and GE, as well as the latest incarnations of Telenet and Tymnet, began to have tougher times in the nineties, the carriers began to take off in managed services. The story of the nineties is really around AT&T, BT, MCI and Sprint leveraging the success of frame relay in building strong managed businesses. AT&T announced in 1995 a new outsourcing division called AT&T Solutions that included a managed and outsourcing part of the business. With frame relay at its peak in growth in the late nineties, the business grew rapidly. Managed routers along with new frame relay implementations provided explosive growth for the carriers. At AT&T, 40% of new frame relay clients chose a managed option. They offered network equipment, transport, implementation services and day two managed support for a single price. Other carriers did the same. A bit of a "private cloud"-type offer that was very successful. MCI and BT's Concert collaboration (which included SHL, a large integrator they acquired) competed heavily with IBM, EDS and AT&T for network, and beyond, outsourcing opportunities. It was a highly competitive world in the pre Internet explosion days, no one had any idea what was in store as the decade wound down.

Of course, one of the biggest events in the nineties was in 1994 when the Internet was opened for commercial business. The impact of this started to be felt later in the decade, but truly impacted managed services in the next decade.

2000's to Present

Most of us know the story in the 2000's. The internet and IP networking exploded. Virtual Private Networks changed the landscape quickly and thoroughly. The period of 1999-2001 saw independent MSP's (managed services providers) of all shapes and sizes start up. By 2001 Gartner had identified almost 100 independent MSP's, many backed by private equity money. 9/11 changed everything everywhere, including the managed services market. With corporate spending being cut back, it was a pretty dark time for the managed services market, along with almost everything else. By 2006 there were less than ten independent MSP's.

One interesting trend in the world of managed support is the move of most enterprises away from the mega-outsourcing contracts of the past. An IBM or HP/EDS running the entire infrastructure, and in some ways acting as a "general contractor", have given way to more selective outsourcing. Clients are looking for deep technical expertise and added value and increasingly are willing to source to different managed vendors, depending on the technology tower. It's a trend that almost brought EDS to its knees and has had a big impact at IBM as well. Both are looking for more 'leverageable' services that will create higher client satisfaction and more profit.

The past several years have seen IT vendors of all types begin to move into managed services - equipment manufacturers of all sizes, resellers, distributors, small and mid-size telcos, integrators of all sizes, India-based offshore providers, among many others - have recognized the need to provide managed / cloud / outsourced infrastructure support. The demand for these services has increased due to client's needs for utility/ opex services, help in running an increasingly complex and often diverse infrastructure, a need to reduce costs and limit headcount, improve performance and reduce the overall risk of running your infrastructure. The boom time for managed services is here, and looking ahead, I certainly see a dynamic environment with many changes and opportunities for vendors who know how to provide high quality managed support at a competitive price, a price which enables them to make a profit and the client to save money.

Chapter 13 - Looking Ahead

While managed services in form or another has been "the" hot thing in IT services off and on for the past thirty years, it has never been quite as hot as it is now. Whether it is called managed services, cloud, private cloud, outsourcing, infrastructure services or some combination/iteration of the above, the trend of companies looking to their IT vendors to help them reduce the cost and improve the performance of their infrastructure is here to stay. From where I sit, I see the demand and desires of companies to leverage these types of services outstripping the true capabilities of vendors today. And part of that is because there simply are less people who know how to effectively run these types of businesses than are needed. I believe anyone who has or will be investing their time to master this craft will be investing wisely. The future in this field is going to be ripe for opportunity.

The world of IT and IT procurement is changing dramatically, and managed services is smack in the middle of it.

One change is the move to what I would call an "IT Utility Model." My spin on this is a bit different than what you read from analysts and in the trades. I choose the words IT Utility Model somewhat carefully. Because I don't see this trend purely as cloud.

What I see happening revolves around three motivations which are beginning to dominate the landscape of IT. One is the desire of clients to spend much less (as close nothing as possible) up front in implementing new technology solutions. The world of spending most of the lifetime cost of an IT solution up front is fading. It puts too much burden and risk on the client and not enough on the vendor. Two, clients are looking to shift the complexity of IT to their vendors. It is starting to become a given, and often a source of frustration, finding someone that can do it to their satisfaction. Three, they want to pay for it in a price times quantity model. I don't often see unrealistic client expectations to have unlimited up and down price flexibility, that puts too much risk back on the vendor and they know that. But they do want to pay for what they use, and it helps them greatly in working with their customers (end users).

There may be more to this shift than these items, and I may not have it totally right, but some derivation of this I believe is happening in a very dramatic way. And I don't think the market it totally getting this, because what is happening out there is not all about the cloud.

> As an aside, I personally believe there is a bit of "over-hype" around public cloud models. I do believe a public cloud model has a large role, and an important one, in this new IT utility model. But the hype in the market, and the spending by IT vendors seemingly everywhere, to me, does not match the reality of the opportunity. The customization required to meet individual client needs, among other things, is too important a factor for public cloud to completely dominate the landscape in the future. Many IT vendors I talk to are not making the profit they need to, some are losing quite a bit of money.

I really do see this IT Utility Model trend as the driving force behind the success Avaya has had in managed services, and Avaya's success in managed services is strong evidence of the speed and ferocity these changes. The world of IT and IT procurement will never be the same. It's not a tidal wave that will impact everything it touches in the next year. It's a tsunami that will take a while to fully take hold, but when it does its impact will hit the entire market in a very big way. I may be a bit over dramatic in describing this trend, but I just feel so strongly about it. And I do believe vendors must step back and invest wisely and aggressively to succeed in this upcoming new world.

There are clear trends occurring today in IT, and the three areas highlighted above are front and center. The positioning of vendor offers needs to be a spectrum of options to meet these requirements - offers including public cloud, private cloud, managed, and financing with optional professional services. Being able to outline your view of what clients are telling the market and reviewing how you will position your utility offers will be critical. I guess overall I would say "this ain't as simple as the cloud" - it is a complex set of new requirements that demand a full set of solutions. Public cloud offers are a part of what vendor's need to provide, but only one part, and only when it truly meets what the client is looking for.

Vendors must be able to provide IT models that include the technology as part of the service. That includes public cloud models, where multiple clients have access to a centralized set of applications and infrastructure enabling attractive price point and flexibility, as well as private cloud models. Private cloud models involve a centralized core set of applications and infrastructure for one client, whether that sits on a client's site or in a vendor's site, with the vendor owning the asset. Some vendors, including some very large ones, are highly averse to this (owning the asset). They need to wake up and figure out a way to have assets sit on their books in a scalable fashion, even if they use a third party. Most large companies will move to this new utility model in a private cloud model, the "one size fits all" approach of public cloud is not going to take over the large enterprise market, that's for sure.

Another key trend is who vendors are going to work with at the client in this new IT Utility Model. Vendors of course typically work with the central IT team, who in turn work with the key business owners. There is no question about the trend of the business owners having more and more power in IT decisions. Often the folks we talk to in IT are only really conduits for what their business owners are driving them towards. It can be a tough relationship between the central IT team and their business owners. These thin, central IT teams are really not in the client relationship business and would actually prefer to make all the decisions. They could use all the help they can get in making their end customers happy. The CIO often gets pushed by the CEO, who gets pushed by his business line management, and the result can be a challenge for the CIO and his/her team. The central IT team lives in fear of upsetting these user groups and is highly motivated to have good "customer satisfaction."

This is not a new trend, but boy is it growing in steam. Your revenue as a vendor is often dependent on the speed of the roll-out and the acceptance of your products by end users in the client's various business units. And often the central IT team isn't doing a very good job of promoting and supporting your products, and they typically don't want you talking to the business group execs directly. Figuring out how to get to the end user decision makers has long been a part of vendor strategies in selling applications. It now needs to be a bigger part of how managed services in its many forms are sold, not only up front but on an ongoing basis. Because the service is procured in a PXQ model, more usage means more revenue you will realize. The more the business unit executives are pleased with the service, the more volume they will procure. Selling to business units up front and supporting them effectively and in a visible manner will be a critical success factor going forward in this market.

Certainly one of the more interesting areas in managed services looking ahead is the workplace area - desktop management and associated help desk support. With more and more business units exerting unprecedented influence on IT decisions, especially related to public cloud alternatives, the CIO and their IT teams have their hands full. Users are bringing their own devices into the workplace and are increasingly using social media at work, for work purposes (mostly, hopefully), At the same time, the IT teams are being pushed to drive lower costs in the workplace, while insuring a highly secure environment. So IT teams feel a strong need to gain more control, while at the same time the actual technology decisions are more and more being dictated to them in one way or another by the business units they support. The CIO and their team have quite a challenge managing this environment going forward, and this will provide strong opportunity for vendors. The workplace management environment for vendors has been a tough one in the recent past. Margins are very tough to come by in an increasingly commoditized environment, and vendors are getting very picky about deals they pursue and even no bidding renewals on current deals.

With CIO's looking to reduce costs and vendors often stuck in a low margin spiral while competing for these deals, the future could appear dull and ultra-commoditized for the workplace. But I think that is going to change fairly rapidly and dramatically. And I believe there is a huge opportunity for vendors who can provide solutions to CIO's to help them navigate through this new workplace world. CIO's must gain control of these hybrid environments cropping up and most likely they won't be able to do it on their own. Looking ahead, I see more and more CIO's successfully arguing that they must have more control in running and supporting their workplace environments, it is getting too out of control. While I see no reduction in business units driving the actual technology decisions (with CIO's often just "implementing" those decisions), I do see CIO's gaining more control in how those technology decisions are rolled out and supported. There are several areas where, if vendors can step up, there is not only strong top line opportunity, but also much improved margins.

Standardization - Nothing too new here, but workplace management vendors must push for standard user profiles, the more the better. Profiles like knowledge worker, power user, light user, etc., will help to keep costs low from a per user cost standpoint for IT departments. Business units will likely have no choice but to accept these type of support scenarios. Mobile management and support is a must, and the quality of vendor solutions here is quite wide now. IT teams need help in order to order, provision, track and manage, which creates excellent value-added opportunity for vendors.

Automation - I really see this as the next wave of efficiency for enterprises and vendors following offshoring. Offshoring is here to stay in one form or another, but I believe much more incremental benefit from this point on lies in increased automation, and those vendors that step up effectively will reap the benefits. Automated end user support, including client portals to provide help, facilitate changes, check status, etc. are an excellent opportunity area. More and more automation to reduce end user calls, enabling them to do "it" themselves, whatever that "it" may be, is a key trend. Integrating social media support for end users also presents excellent opportunity. The basics of remote management are becoming more sophisticated, and better and better preventative monitoring support will continue to present opportunity to predict potential problem areas and act. Workplace management vendors that bring strong intellectual property around automation to the table will be in excellent position to succeed.

Virtualization - We are just scratching the surface on the benefits of virtualization, and in the workplace, HVDS (hosted virtual desktop services) I think are the wave of the future, though it may take some time to get there. These services, where clients have their desktop virtualized centrally on core servers vs. on their desktop, do in some ways inhibit some user flexibility, but the advantages in control, security, and efficiency are tangible. It won't be for everyone, certainly not for high travel users, but over time I see HVDS playing a pretty strong role in the workplace. I don't think it necessarily will reduce costs for IT teams, but the benefits in security and control will be highly attractive.

The low cost, highly commoditized world of workplace / desktop / help-desk management is going to change. The CIO and team, with public cloud, private cloud, BYOD and various hybrid environments cropping up, are going to increasingly convince their businesses this must get under control or will simply be asked to do it. So while their role in specific technology decisions will likely reduce a bit, their role in making it all work under increasingly difficult environments will increase. And this change I feel creates excellent opportunity for workplace management vendors to create true value (and good profit, finally) in helping IT teams manage these hybrid environments

Certainly those of us who have been in managed services for a while have a lot to look forward to in the coming years. It seems as though many trends in IT are coming together to have managed services, outsourcing, cloud - whatever label you want to put on it - move to the forefront of the next big wave in IT.

Chapter 14 - The Agony of Managed Services

Managed Services is a great business. I've been telling myself that for thirty years, and, sometimes, it can be a little hard to believe. Running a managed or cloud business when you are a telephone company, an equipment manufacturer, a technology distributor or reseller, or any other IT company (other than a pure play managed provider), can be a very frustrating business to run and be a part of. You are often, or should I say typically, working for people running the company who are there because they are the best at what they do. And what they are best at is most definitely not managed services.

I could write a whole book about my adventures working with senior executives at companies I have been at, trying to get them to support managed services. It would be a story filled with tales of agonizing debates, incredible lack of foresight, full faith and trust, complete lack of faith and trust, bouts of confusion, incredibly bad decisions, remarkably insightful ones and everything in between. I have no intention of reliving or wallowing in those memories. I have really enjoyed building and running managed businesses, but I have really not enjoyed the continual struggle it has been to keep the forces above me from making bonehead moves to derail the business. Everyone I know who has spent any time in a managed business working for an IT company shares this bond, this badge of honor, that we all would like to forget and wish would go away. But that's not happening.

To run a successful managed business inside a company that fundamentally does something else is not easy. But, of course, it is a critical success factor for many managed businesses.

And honestly, while more and more companies are now focused on having a managed business, I don't see it getting any easier. Describing the benefits and convincing those running the company you should be in managed services is relatively easy. Making the tough decisions required on a day by day basis that support having a successful managed business is often a very different story.

It is those day-by-day decisions which happen regularly that need support from the top of the company - everything from insuring the compensation plan for your sales team incents managed services in addition to your core offering, making the right investment in your platform and tools to drive success, keeping your managed business organizationally focused on managed services vs. splitting it up to gain small efficiencies (and dilute your effectiveness), and having your finance department understand the intricacies of an annuity vs. one time revenue business. The potential pitfalls are many, and just when you have things well set up for success, there inevitably will be curves thrown your way that you will have to overcome to be successful.

I don't really see any fool proof method to insure you have the right support. I do think clearly communicating the keys to you being successful to any leader of your business is a must. One thing I have learned is to never assume some of the basics you take for granted in running a successful managed business. You really need to keep reminding leadership of the keys things you need to be successful. Those keys to me are pretty much outlined in the ten plus one commandments at the end of this book. Any managed business is going to need support of those items from the top of the business, and reminding leadership every reasonable chance you get is worth the effort.

I'm a big New York Giants football fan, and my favorite coach was Bill Parcells. I remember him in an interview a while back once say that the biggest key to success for him was having an owner who fully supported him . . . . that when that bond is not there, it's very tough to be successful. The owners he worked for became successful and rich having nothing to do with football, and now they are the ultimate decision makers. I recall thinking at the time that can't be *the* key, this is a game played on the field, you need great players and a great game plan. But as the years went by I have started to get what he was saying more and more. Bill Parcells had a tough business to run, being successful in the NFL is brutal, and without the right structure, support and decision making at the top, you won't get there. I see similarities to managed services more than most businesses in IT, because it is such a niche business with so little understanding of what it takes to be successful.

In a way, it comes down to the benefit your company sees by being in managed services, and what you must have to be successful in having that managed business within your company. It's important to emphasize and remind your leadership why managed services is so important to your company, and I think it's good to put it in offensive and defensive terms. Offensively, managed services is a great opportunity, it's a fast growing market, clients are increasingly moving that way, and once successful, its predictable revenue and profit are huge benefits the business will realize. In addition, as we discussed, 60% of managed services contracts lead to an upsell and 85-90% of all future technology decisions are dictated by a client's managed services provider.

When things get tough, however, you can find leadership thinking while all that's true, I may be able to realize more revenue and/or profit by putting my support in another area. That's where defensive motivations are important. The focus needs to be on your clients, and if they are increasingly looking to managed solutions, whether a full opex model or more basic managed model, you must have a competitive solution in place or your competitors will. And as we all know, once someone else is running the infrastructure of your client, you have lost control. The managed services bond between client and vendor is very tough to break, and the place the vendor has at the table with the client on whatever happens with the part of the infrastructure they run is significant. To continually insure you are well positioned with your clients, it's critical to have a competitive and successful managed business. And to have that successful managed business, there are things you need that must be continually reinforced to your leadership.

When you have a Wellington Mara as your owner (the Giants owner when Bill Parcells won his two Super Bowls) and you get the support you need, you are a long way towards having the kind of business you and your leadership want. It's an art and science to keep insuring you have that support, and not assuming anything is probably the best mantra to have in this regard.

Chapter 15 - Managed Services Ten (plus one) Commandments

1) Managed Services is a high fixed cost, low variable cost business which needs to be run that way.

2) A dedicated, motivated and 'managed services-experienced' overlay sales force is required to generate demand and close business when selling through a front line sales force.

3) You need to have a relentless pursuit of asking the right open-ended questions during a managed services sale and reacting appropriately - getting to the point where you are jointly planning how to reduce client cost and improve their performance, while always on the lookout for a "compelling event," will inevitably lead to a successful sale.

4) Your core managed services offer must have tangible market differentiation that is validated externally (clients and analysts). Its structure should drive your delivery processes.

5) Your managed platform is the foundation of the service you will be providing. Your documented processes should determine the functionality needed in the platform, and the platform should in turn drive your process execution.

6) Most of your managed services delivery team will need to "follow the process" with a great attitude and client

focus. There are only so many high end "technical cowboys" you should be depending on to deliver quality support.

7) The first 90-180 days of implementing a client on to your managed service will determine with a high degree of probability the success of that engagement over the life of the contract.

8) Small, custom deals kill. While large custom deals, done correctly, can and often do, ignite a managed services business.

9) The leader of your delivery team must be highly motivated to grow the business and increase bottom line profit each and every day.

10) Managed services requires patience, it takes a long time to build a strong, growing profitable business, and once you get there, a long time to screw it up.

"Plus One": It is critical to have a detailed, unit-based strategic operating plan with an obsessive and relentless focus on business growth and cost efficiency metrics.

Chapter 16 - Concluding Thoughts

Those of you who have read this entire book, I appreciate your interest and endurance! I recognize a lot of it is rather detailed and dry, but I hope as you finish the book you feel intuitively more confident and energized going forward and running your managed services business.

Managed Services must be done right from the bottom up. When a strong vision of managed services from a value proposition, differentiation and go to market model standpoint meets an equally strong foundation of execution excellence that delivers the quality, function and efficiency needed, great things can and do happen.

Companies are moving to a utility model for IT, a phenomenon that is not turning back. That fact, along with increasing technology complexity, will fuel an unprecedented and long lasting growth in managed, outsourcing and cloud services. And the leaders in the next generation of managed services will be those who have the experience and operational focus through all aspects of their business to make it work in an optimal fashion.

Though managed services has been around for a long time, it still is, in my opinion, an embryonic market, ripe for the next generation of great companies and leaders to emerge. You may be the person to take you and your company there. I hope to meet, work with and learn from as many of you as possible, as I have great confidence this next generation of great managed services companies and leaders will far exceed anything we have seen to this point. I for one am really looking forward to the future of this little niche IT services area that is now fully ready for prime time.

About the Author

Ed Nalbandian is a veteran of over thirty years in the managed services business, playing a variety of roles and running businesses of many sizes.

He spent his first thirteen years of his career in IBM's managed services business, starting in 1982 with IBM's new "Information Network" division. He parlayed both sales and product management roles to become a key executive who, within a few short years, helped to transform it into the IBM Global Network, an industry leader in managed services.

He was recruited by AT&T in 1995 to start and lead Managed Network Solutions, a new business AT&T was creating. Under his leadership, the unit achieved unprecedented growth and profitability, and was recognized as the industry leader from a market share and client satisfaction standpoint. By 1999, the organization was over $500m with over 1,400 employees.

He started his own managed and professional services company called AimNet Solutions in 2000. He grew it from its small beginnings to become an industry leader in remote managed services, with over 600 clients and 150 employees, as well as broad recognition as a leading independent Managed Services Provider (MSP). He was a finalist for the Ernst and Young "Entrepreneur of the Year" in 2003.

He sold the business to Cognizant in 2006, where he became the head of their IT Infrastructure Services business. At Cognizant he engineered a major restructuring of the business and drove 140% increase in revenue over the two years and led analysts to recognize the company as a tier one infrastructure provider.

He joined Avaya in 2008 to run Avaya's managed services business called Avaya Operations Services. He led a dramatic transformation of the business, taking a declining revenue stream to a robust 20%+ growth business. Avaya's turnaround in managed services was fueled by its innovative "private cloud" outsourcing model based on a global delivery model and platform infrastructure. The business won the Frost and Sullivan North America Market Leadership award in 2011. He is a founding member of the TSIA Managed Services Board.

He is currently the President of Enabling Managed Services, L.L.C., a consulting company focused on helping managed services vendors drive successful top and bottom line initiatives.

Mr. Nalbandian is a 1982 graduate of the University of North Carolina, Chapel Hill with a BA in Economics. He has been married for twenty four years to his wife Tracy and has two children, Kellie, 20 and Eric, 16.

63483131R00132

Made in the USA
Middletown, DE
26 August 2019